Interpretation of Green Mine Evaluation Index

Jiushuai Deng · Suping Peng · Liang Wang ·
Yinli Bi · Jun Yao · Qiongjie Wang

Interpretation of Green Mine Evaluation Index

Science Press
Beijing

Springer

Jiushuai Deng
China University of Mining
and Technology-Beijing
Beijing, China

Suping Peng
China University of Mining
and Technology-Beijing
Beijing, China

Liang Wang
Zhongguancun Green Mine Industry
Alliance
Beijing, China

Yinli Bi
China University of Mining
and Technology-Beijing
Beijing, China

Jun Yao
China University of Geosciences
Beijing, China

Qiongjie Wang
China Mining News
Beijing, China

ISBN 978-981-16-5432-9 ISBN 978-981-16-5433-6 (eBook)
https://doi.org/10.1007/978-981-16-5433-6

Jointly published with Science Press
The print edition is not for sale in China (Mainland). Customers from China (Mainland) please order the print book from: Science Press.

Translation from the Chinese language edition: *Interpretation of Green Mine Evaluation Index Clauses* by Jiushuai Deng, et al., © Science Press 2020. Published by Science Press. All Rights Reserved.

This Springer imprint is published by the registered company Springer Nature Singapore Pte Ltd.
The registered company address is: 152 Beach Road, #21-01/04 Gateway East, Singapore 189721, Singapore

Writing Committee

Director:

Min Wang

Vice Directors:

Runcang Yu, Yinao Su, Suping Peng, Jingwen Mao, Hongpu Kang, Shaohui Wu, Fengjun Guan, Suichen Liu, Xingbi Bai, Ansheng Feng, Renyi Chen, Zesong Hu, Laping Zhao, Youyi Hu

Committee Members (Sort by strokes of the last name):

Shuaidong Wu, Lan Wang, Liang Wang, Shike Wang, Qiongjie Wang, Jing Wen, Shuming Wen, Jiushuai Deng, Fangshou Bao, Sheng Shen, Yuping Shi, Shujuan Dai, Liang Feng, Shiwen Lyu, Ruijun Zhu, Wenguang Qiao, Li Liu, Wenyu Liu, Lilong Liu, Kebin Liu, Jingyong Liu, Peng Guan, Jiaxuan Tang, Yongchao Sun, Genjie Du, Xiaodong Yang, Gang Li, Pu Li, Jiancheng Li, Yanqin Li, Dexian Li, Jinling Shi, Shangkun Wu, Yunchuan Wang, Deping Zhang, Yunhai Zhang, Changqing Zhang, Shixin Zhang, Junying Zhang, Guangfu Chen, Limin Fan, Xiao Zhi, Xin Zhou, Ziqiang Zhou, Huali Hou, Junliang Hou, jun Yao, Junhong Yuan, Mingqiang Xia, Shanqin Ni, Zhongmin Gao, Guojun Huang, Jiagang Huang, dongtao Chu, Aishu Wen, Jun Jin, Jie Pei, Jinhua Tan, Guotai Pan, Zengjun Wei

Authors: Jiushuai Deng, Suping Peng, Liang Wang, Yinli Bi, Jun Yao, Qiongjie Wang

Participants:

Xi Lu, Pu Li, Haijun Wu, Shanyong Liu, Wei Zhao

Preface

"Lucid waters and lush mountains are invaluable assets", "Respect, conform to and protect nature", and "Green, circular and low-carbon development" are the basic concepts of Xi Jinping's Thought on Ecological Civilization, and also the basic concepts that should be followed in the construction of green mines. In May 2020, *Notice of the General Office of the Ministry of Natural Resources on Accomplishing the Selection of Green Mines in 2020* (Natural Resources Office (2020) No. 839) issued by the General Office of Ministry of Natural Resources of the People's Republic of China (Hereinafter referred to as Ministry of Natural Resources) laid out and standardized the green mines selection in detail, and set the tone for evaluating the construction level of green mines. With the publication of the *Green Mine Evaluation Index* and *Requirements for Third-Party Evaluation of Green Mine Selection*, the nationally unified evaluation criteria and related regulations for selecting green mines were officially issued, and the selection of green mines in each province officially began. *Green Mine Evaluation Index* is based on the Green Mine Construction Specification's specific requirements for nine industries, combining the actual situation of the mine and widely solicited opinions from all walks of life. It has the following characteristics.

Firstly, the evaluation indicators reflect the connotation of green mine construction. The evaluation indicators focus on the core concept of the coordination between resource development and ecological protection, contribute to the mines' green development and the industrials' transformation and upgrading, and boost mining enterprises' development toward the high-quality goal of "sustainable, harmonious and efficient".

Secondly, the evaluation indicators strive to have a concise description and accurate meaning. Considering the wide range of personnel using the evaluation indicators and the large differences in professional knowledge, the evaluation indicators in mining technical requirements, mining face quality requirements, beneficiation, and mineral processing technology shall be described as concisely and accurately as possible to avoid ambiguity. Sand, cement, and other mines differ significantly from other mines in terms of comprehensive resource utilization due to different resource occurrence conditions, and they are also clearly distinguished.

Thirdly, the evaluation indicators reflect the guidance of green mines construction. The evaluation indicators increase the management requirements of green mine technical equipment, improve the green mines management system's construction content, and strengthen scientific and technological innovation's specific connotation. These changes reflect the focus and guidance of green mines construction and clarify the direction for the next step to promote green mines construction.

Fourthly, the evaluation indicators indicate the path of green mines construction. Many people think of green mine construction as a concept that mainly focuses on environmental governance and greening of mining areas; however, they are confused about building green mines. The evaluation indicators tell mining enterprises to sort out specific and practical work, such as mining equipment, dust production points, and noise production points so that mining enterprises can grasp the path and practical methods of building green mines.

The Zhongguancun Green Mine Industry Alliance associating relevant units has compiled Interpretation of Green Mine Evaluation Index to enable green mine workers to fully understand the evaluation indicators and accurately apply them to green mines construction, self-evaluation, third-party evaluation, and on-site verification. The book based on the actual situation of green mine construction by mining enterprises illustrates the relevant concepts, relevant laws, policies, implementation measures, inspection points, and materials that enterprises should provide in the green mine evaluation standards. It can be used by mining enterprises, green mine consulting service agencies, third-party assessment agencies, and mine administration managers for green mine construction and assessment.

Thanks to the guidance of Zhongguancun Green Mine Industry Alliance, China University of Mining and Technology (Beijing), China Mining News, China University of Geosciences (Beijing), Weifang Tianjie Environmental Technology Co., Ltd., Beijing Gaozhuo Dongsheng Technology Co., Ltd., Beijing Lvhaishengyuan Certification Service Co., Ltd., Sichuan Mineral Resources Research Center of the Chengdu University of Technology, etc. This book refers to many domestic and foreign documents, so if there are any omissions in the notes, I would like to express my apologies. Now, I would like to express my sincere thanks to the mentioned and unmentioned authors and publishers!

Despite the author's best efforts, there are inevitably some unsatisfactory points due to the limited level. I adjure readers to criticize and correct. Let us work together to promote the mining enterprises' green development and industrial transformation and upgrading.

Beijing, China

Jiushuai Deng
Suping Peng
Liang Wang
Yinli Bi
Jun Yao
Qiongjie Wang

Description

The content of this book includes an introduction, prerequisites for green mine selection, green mine construction score sheet, related knowledge, as well as introductions of mining area environment, resource development methods, comprehensive utilization of resources, energy conservation and emission reduction, technological innovation and intelligent mines, enterprise management and enterprise image, and other specific green mine evaluation sub-items.

This book can be used as a reference book for administrators, engineering and technical personnel, related professional teachers and students of mining enterprises, government departments, consulting services and evaluation agencies, colleges, and secondary professional schools.

Contents

Chapter 1
Introduction

In 2018, the Ministry of Natural Resources issued nine industry standards, including *Non-metallic Mine Industry Green Mine Construction Specification*. From the development concept of innovation, coordination, greening, openness, and sharing, the standards clarify the basic connotations of green mines. Green mines refer to the implementation of scientific and orderly mining in the whole process of mineral resource development to control the mining area's disturbance and surrounding ecological environment within a controllable range and realize the ecological environment in mining areas, scientific mining methods, efficient resources utilization, enterprise management's standardization, and mining communities' harmony [1]. Therefore, green mines are mines that protect the ecological environment to the maximum extent in resource development; the supporting facilities, technical equipment, and management level of the mining area can ensure the continuous, harmonious, and efficient operation of the mine. It is a lively community of mountains, rivers, forests, fields, lakes, grasses, mineral and gas, a combination of ecology and industry, a systematic whole-process thinking, a system construction, an important part of ecological civilization's construction, the practice of ecological civilization construction in the mining industry, and a choice of paths [2–6].

Green mines, combined with the current new situation and new development requirements, shall be a new mode of mine construction and operation guided by the strategy of ecological civilization to develop a green economy and harmonize resource benefits, ecological benefits, and economic and social benefits. Green mines shall be based on the premise of operating mines in accordance with the law, standardizing management and safe production, focused on efficient cultivation, environmental protection, energy-saving, emission reduction, and community harmony, guaranteed by technological innovation [7].

Green mines construction has become a national strategy and action from advocating the concept of green development, gathering the consensus on industrial development, actively exploring and practicing all over the country [8, 9]. It has played an important role in transforming resource utilization methods, improving resource

J. Deng et al., *Interpretation of Green Mine Evaluation Index*,
https://doi.org/10.1007/978-981-16-5433-6_1

utilization efficiency, coordinating the relationship between resource development and ecological protection, and promoting enterprises' social responsibilities [10–12].

In 2003, the Third Plenary Session of the 16th Central Committee of the Communist Party of China put forward "establish a comprehensive, coordinated and sustainable development concept, and promote the overall development of economy, society and people" for the first time, and the concept of the green mine came into being.

In 2005, the green-mine concept was put forward in the *Notice of Zhejiang Provincial Department of Land and Resources on Carrying out the Pilot Work for Establishing Provincial Green Mines* for the first time.

In December 2008, the State Council formally promulgated the *National Mineral Resources Plan (2008–2015)*, which put forward the requirements of developing a "green mining industry" and set the overall goal of "Establish green mines pattern basically by 2020."

In August 2010, the Ministry of Land and Resources issued the *Guiding Opinions of the Ministry of Land and Resources on Implementing the National Mineral Resources Planning and Development of Green Mining and Green Mines Construction*, which put forward precise requirements for green-mine construction.

In March 2011, the Ministry of Land and Resources announced the first batch of national green mine pilot list, marking that China's green mines construction has entered the pilot demonstration stage. From 2012 to 2014, the Ministry of Land and Resources successively announced the second, third, and fourth batches of pilot units, totaling 661.

In October 2015, "we must adhere to the basic state policy of conserving resources and protecting the environment and sustainable development, accelerate the construction of a resource-saving and environment-friendly society, and form a new pattern of modernization featuring harmonious development between man and nature to insist on green development." It was proposed in the communique of the Fifth Plenary Session of the 18th CPC Central Committee.

In 2017, *Implementation Opinions on Accelerating the Green Mines Construction (Land and resources regulation [2017]No. 4)* jointly issued by six departments including the Ministry of Land and Resources proposed "Establish thousands of science and technology-led, innovation-driven green mines model, implement one hundred green exploration project demonstrations, build more than 50 green mining development demonstration zones, and form a batch of new models, new mechanisms, and new systems that can be copied and promoted" in the overall goal of green mines [13]. It clearly stated that the green mines' characteristics are "Technology-led and innovation-driven" and highlighted the critical role of scientific and technological innovation in green mines construction.

Meanwhile, the "*Implementation Opinions on Accelerating the Green Mines Construction*" puts forward the requirements of clarifying mine environment's appearance, development and utilization methods, resource conservation and intensive utilization, modern mines construction, mines harmony, enterprises' cultural image, and other green mine construction assessment indicators, and makes exact requirements for the key work of green mines.

In 2018, the Ministry of Natural Resources issued the *Green Mine Construction Specification* for nine industries, including the chemical industry, non-metallic ore industry, and gold industry. Inner Mongolia, Henan, Shaanxi, Guangxi, Jiangxi have successively issued provincial-level green mines construction's specifications and construction plans. Also, some provincial schemes are open for comment.

In May 2020, the General Office of the Ministry of Natural Resources issued *Notice of the General Office of the Ministry of Natural Resources on Carrying out the Selection of Green Mines in 2020* (Natural Resources Office (2020) No. 839). In June 2020, the Ministry of Natural Resources issued *Work Requirements for the Third-Party Evaluation of Green Mines Selection* to unify the evaluation indicators standards, standardize the third-party evaluation work, and clarify that "The evaluation indicators scoring table includes 100 items, with a total score of 1,000. The construction level of green mines is scored from six categories: mining area environment, resource development methods, comprehensive resource utilization, energy conservation and emission reduction, science and technology innovation and intelligent mines, enterprises management, and enterprises image." It is the first time for the country to issue evaluation indicators related to green mines, which makes up for the lack of adequate evaluation indicators in the definition and evaluation of green mines. With the continuous promulgation of green mine policies, more and more enterprises actively respond and strive to build green mines [14, 15]. The release of the green mine evaluation indicators can provide the basis for the judgment and definition of green mines and can be used to measure and guide the standardization and management of green mine construction and mining enterprises' development. Meanwhile, it will also promote new mining technologies, new models, and new forms of business.

However, understanding and implementing these policies, standards, and requirements need in-depth study and analysis of green mines evaluation indicators.

From the aspects of difficulty, depth, and content of green mines evaluation indicators, the green mines' evaluation indicators can be interpreted.

Firstly, from the perspective of indicators' difficulty, the difficulty of completing indicators includes three categories: difficult, ordinary, and easy for enterprises creating green mines.

"Difficult" means that mining enterprises need to invest more workforce, material resources, and financial resources to improve their work, such as "77 Intelligent Working Face or Unmanned Mining Cart System" (the number indicates the serial number of the three-level index in the evaluation index, the same below), "26 Environmental Management System Certification", and "67 Science and Technology Awards." These indicators do not belong to the "difficult" types if the original design scheme meets the indicators' requirements. In contrast, the original design scheme needs to be completed with the help of mining technical transformation if it does not meet the index requirements.

"General" refers to the index requirements that mining enterprises can achieve with a little effort, such as "56 Domestic Sewage Discharge" and "75 Remote Video Monitoring System."

"Easy" refers to the routine work carried out by mines in normal production, such as "1 Functional Area", "95 Employees' Physical Examination", "100 Information Announcement," and so on. "Easy" includes part of the "sending points" indicators, for example, relating to the comprehensive utilization of co-associated resources and carrying out the project of comprehensive utilization of co-associated resources could score. Those that cannot exploit the associated minerals, for the time being, shall be considered as non-related items. Taking effective protection measures for these co-associated resources can also be scored. It reflects the importance of co-associated resource development, which is a kind of guidance.

Among the 100 green mine evaluation indicators, most of them belong to the "general" type, and mining enterprises can quickly achieve the target requirements as long as they strive to improve.

Secondly, from the perspective of indicators depth, the requirements of the indicator can be divided into four levels: "Is there any," "Is it right," "Is it done," and "Does it work."

"Is there any" means whether there is a certain content required by the indicators, such as "72 Intelligent Mine Construction Plan", "51 List of Major Dust Production Sites", and "64 Technical R&D Teams."

"Is it right" examines the relevant requirements of "Is there any" and whether the specific relevant content is correct. The evaluation goes further compared with "Is there any," for example, "89 Functional Area Management System", "4 Production District signs" and so on have set requirements for relevant specific content.

The key to consider about "Is it done" is whether the mining enterprises have already implemented relevant requirements, such as "9 Main Road conditions", "10 Road Cleanliness Conditions" and "11 Cleaning Conditions of Mining Area", these are all requirements that mining enterprises must do.

"Does it work" focuses on whether to play the indicators specified function role, for example, "22 Governance Requirements", "23 Land Use Function Requirements", and "24 Ecological Function Requirements" have proposed "shall meet," and "should comply with" and other effective regulations.

"Is there any" and "Is it right" correspond to the fundamentals, while "Is it done" and "Does it work" require to be done. The indicators shall reflect the level of "Does it work" if you want to truly identify and evaluate high-quality green mines. Many evaluation indicators now stay at the level of "is there any" and "Is it right," indicating that the indicators' requirements are gradual.

Thirdly, from the perspective of the indicators' content, the indicators' requirements include "mechanisms," "programs," "regulations," and "processes." The indicators try to evaluate the green mines' construction level from the aspects of relevant mechanisms' establishment and operation, programs' formulation or implementation, regulations' implementation, or process management.

"Mechanism" refers to the structural relationship and operation mode among the elements and the system's construction. Greening mechanism, emergency guarantee mechanism, technological innovation mechanism, and green mine management system all belong to mechanisms. Mechanism indicators involve laws and regulations, system formulation, assessment standards, plans, organization and personnel, training and publicity, operation status, and operation results.

"Plan" mainly refers to implementing the *Preliminary Design Plan, Mine Geological Environmental Protection and Land Reclamation Plan, Intelligent Mine Construction Plan,* and other relevant plans. Such indicators are mainly required from the program's content or design, the program's implementation, and the effect of the situation.

"Regulations" mainly refer to the provision of relevant standards, such as *Specifications for Design of Factory and Mine Roads* (GBJ 22-1987), *General Industrial Solid Waste Storage and Site Pollution Disposal Control Standards* (GB 18599-2001), and other regulations and requirements. Such indicators are mainly evaluated from aspects such as stipulated content requirements, evaluation standards, program implementation, and effects.

"Process" mainly refers to some requirements in the process of mines' production and operation, such as fixed management and visual management. Process indicators mainly include specific requirements, content, location, and other requirements based on local conditions.

Green mines construction involves many policies and standards and a wide range, and relevant industry departments supervise each aspect of the requirements. Building green mines can consider the concrete content of the following five aspects.

Firstly, change the disorganized into the organized.

It mainly considers how to change the unorganized emissions of dust and waste gas into organized emissions to meet environmental protection's requirements. Formulating reasonable improvement measures is the key to change the unorganized into the organized.

Secondly, change the uncontrollable into the controllable.

It mainly includes applying green mining technology to keep the geological environment in a controllable state, filling mining to reduce surface subsidence and rock displacement, smooth blasting, millisecond blasting, and pre-splitting blasting to protect the geological environment of open-pit mines [16]. Through the wastes stacking, classification, environmental protection, operation management, shutdown, and monitoring to control wastes; through sound absorption, sound insulation, vibration reduction, vibration isolation, and other aspects to control noise emissions.

Thirdly, change the irregular into the regular.

There are three cases. Form a closed-loop waste management mechanism, greening mechanism, technological innovation mechanism, environmental monitoring mechanism, and other relevant mechanisms or systems from eight aspects, including sorting out laws and regulations to comply with, institutions, assessment standards, work plans, organization and personnel, training and publicity, operation process and operation effect. Ensure the mining design scheme, the beneficiation design scheme, the intelligent mine construction scheme, the sewage treatment scheme form a closed-loop, effective and standardized production process from aspects including sorting out the plan, and the implementation and effect of the plan. Investigate the implementation effect of the mining layout, household garbage collection, classification and treatment, and the information announcement from the provision of relevant standards, the situation's implementation, and results achieved.

Fourthly, change the non-standard into the standard.

Visual management, fixed management, and other management methods are used to standardize the production process, thus achieving efficiency, environmental protection, safety, and high quality.

Fifthly, change the incomplete into the complete.

It includes three aspects of the system, facility, and management's improvement. For example, the road after hardening also needs to repair drainage ditch and roadside greening.

Green mines have become an important platform and starting point for transforming mining development methods and improving the mining industry's overall image, the mining industry's healthy and sustainable development. The green mine evaluation indexes are an important basis for the green mine list's management and supervision and an important step for establishing a scientific and standardized evaluation index system. It is of great significance for more mining enterprises to be included in the national green mine list [17].

The green mines' construction process is to check and fill the gaps following relevant standards and evaluation indicators, and requires mining enterprises to regard green mine construction as daily work. The green mines' construction process shall be organically integrated with mine management to discover problems and formulate plans for continuous improvement [18]. The improvement of facilities, processes, and equipment by mining enterprises shall be combined with continuous technological transformation and innovation to gradually improve green mines' construction and realize the mining industry's green development.

References

1. Liu J (2014) The connotation of green mine and it's system structure. China Min Mag 2:51–54
2. Ju, Jianhua (2014) The green future of China's mining development. Earth 11:32–33
3. Ju, Jianhua (2020) Building the support system of green mine construction in China. China Min Mag 29(1):13–15
4. Ju, Jianhua, Huang X, Xue Y (2018) Thoughts of mineral resources conservation and comprehensive utilization in China in new era. China Min Mag 27(1):1–5
5. Ju, Jianhua, Qiang H (2017) The trend and direction of green development of the mining industry in China. China Min Mag 26(2):7–12
6. Ju, Jianhua, Wang Q, Chen J (2019) Study on the high quality development of China mining industry in the new era. China Min Mag 28(1):1–7
7. Hou H, Qiang H, Chen L (2018) Study on the green development and high quality development of mining industry in the new era. Nat Resour Econ China 31(8):4–10
8. Wu, Shangkun, Yu, Huali Hou, Dong. (2019) Thoughts on the government's preparation of green mining development plan. Nat Resour Econ China 32(4):16–19
9. Zhang Y, Hou H, Nie B (2016) Devoting major efforts to the development of environmentally friendly mining industry to promote sustainable mining development. Nat Resour Econ China 29(11):15, 27–29

10. Hou H, Wu S, Song M (2019) Seeking sustainable development of mining industry. China Natural Resources News, 006
11. Liu X, Hou H, Guo D (2020) The connotation and effectiveness of China's green mine construction in the new era. China Min News, 002
12. Meng X, Hou H, Wu, Shangkun (2018) Discussion on "green" thoughts in the development of mining industry. China Min Mag 27(8):85–87
13. Hou H, Wu S, Jiang F, Dong Y, Song D (2019) Thoughts on green mine construction planning in the new era. China Min Mag 28(7):81–85, 93
14. Wu S, Hou H (2017) Let green become the main keynote of mining development. China Natural Resources News, 005
15. http://www.gov.cn/xinwen/2018-08/01/content_5310980.htm
16. Chen G (2008) Opencast mining technology. Metallurgical Industry Press, Beijing
17. Zhou J, Ju, Jianhua, Yao Y (2013) Exploration on taking Tongling city as the center to construct national mining economic test area. China Min Mag 22(8):5–8
18. Zhang F (2009) Study on ecological environment compensation mechanism based on negative effect in mining. China Ind Econ 12:5–15

Chapter 2
Prerequisites for Green-Mine Selection

There are five prerequisites for green mines construction, which mainly reflect that mining enterprises shall abide by national laws and regulations and relevant industrial policies and run mines according to the law (See Table 1).

In addition to the above five veto conditions, there are also the following relevant requirements when selecting the green mines.

1. Relevant laws and regulations
   According to the requirements of *Notice of the State Council on Printing and Distributing the Reform Plan of the Mineral Resources Royalty System* (Promulgated by The State Council [2017] No. 29), mining enterprises shall pay mineral resources royalty following regulations.
2. Related policies of the Ministry of Natural Resources
   Whether mining is implemented by Mineral Resources Exploitation and Utilization Plan.
   The effectiveness and performance of the Mine Geological Environmental Protection and Land Reclamation Plan.
   Whether the mine geological environment governance and restoration fund (including the land reclamation fund) is deposited in the relevant account on time.
3. Other industrial policies
   (1) Industrial policy requirements of relevant ministries. For example, *Work Plan for Classified Disposal of Coal Mines Below 300,000 Tons per Year* jointly issued by the National Development and Reform Commission, Ministry of Finance, Ministry of Natural Resources, Ministry of Ecology and Environment, National Energy Administration, and National Coal Mine Safety Administration puts forward the working goal to reduce the number of coal mines under 300,000 tons per year to less than 800 throughout the country by the end of 2021 in 3 years, coal mines under 300,000 tons per year in North China and Northwest China (excluding Southern Xinjiang) have been eliminated. The number of coal mines

J. Deng et al., *Interpretation of Green Mine Evaluation Index*,
https://doi.org/10.1007/978-981-16-5433-6_2

Table 1 Prerequisites for green mines selection

Prerequisite	Specific situation
The licenses are legal and valid	*Business License*, *Mining License*, and *Security Acceptance Letter* are legal and valid
No administrative penalty within 3 years	In the past 3 years (3 years prior to the issuance of this selection notice), no administrative punishment has been imposed by the departments of natural resources and ecological environment, or the punishment has been rectified (certificates issued by relevant administrative departments), and no major safety and environmental protection accidents have occurred In the past 3 years (the first 3 years from the date of issuance of this selection notice), no administrative penalties have been imposed by natural resources and ecological environment departments, or penalties have been rectified (certificate issued by relevant management departments), and no major safety and environmental accidents have occurred
Directory of abnormal mining owners	During the mine participation selection period, the mining owners shall publicize the mining owners' exploration and mining information, and the mining owners shall not be included in the abnormal directory of the mining owners' exploration and mining information disclosure system
Mining requirements	The mine is in normal operation, and the recoverable period of the remaining reserves (according to the annual report of reserves) shall not be less than 3 years
Minefield	The minefield does not involve any kind of natural reserve

Data source Letter on Printing and Distributing "Green Mines Evaluation Indicators" and "Requirements for Third-Party Evaluation of Green Mine Selection (Natural Mineral Resources Guarantee Letter [2020] No. 28) Accessory 1-Evaluation Indicators of Green Mine Construction

below 300,000 tons per year will be reduced by more than 50% from the end of 2018 in other regions in principle. In this way, coal mines below 300,000 tons per year in North China and Northwest China (excluding Southern Xinjiang) are not suitable for selection.

(2) Relevant requirements of provinces, cities, and counties. *Three-year Action Plan for the Special Improvement of Production Safety in Shanxi Province* proposes that all coal mines with a production capacity of fewer than 600,000 tons shall be eliminated before the end of 2020. In this way, coal mines in Shanxi with an annual output of fewer than 600,000 tons shall not participate in the selection.

The followings are analyses of the prerequisites for green mine selection.

I. The licenses are legal and valid

Requirements: *Business Licenses, Mining Licenses, and Security Acceptance Letters* are legal and valid.

[Interpretation] This article stipulates that the most basic production and operation certificates must be legal and valid.

This requirement is the rejection requirement of the evaluation indicators determined according to the requirements of "4.1 Mining enterprises shall abide by national laws and regulations and relevant industrial policies, and conduct mines following the law" in *Green Mine Construction Specification* (If there is no other description, *Green Mine Construction Specification* refers to the relevant specifications of nine industries, in which case the requirements of nine industries are the same and will not be described separately below).

Business License is a certificate issued by the administrative department for industry and commerce to industrial and commercial enterprises and self-employed individuals for permission to engage in specific production and business activities. The State Administration shall uniformly stipulate the format for Industry and Commerce. Industrial and commercial enterprises or self-employed individuals without business licenses cannot start a business, engrave official seals, sign contracts, register trademarks, or publish advertisements, and banks will not open accounts.

A mining License is a legal certificate for the mining owners to exercise the right to exploit mineral resources. It is a license issued by the mining registration management authority to grant the application for the mining right to exploit mineral resources. Mining rights licenses shall be uniformly printed by the State Council's natural resources department and issued by the natural resources department at all levels in accordance with the legal authority. The mining license's main contents include mining enterprises' names, economic nature, main minerals, associated minerals, three-dimensional range of mining area, and valid period. Mining licenses shall not be bought or sold, altered, or lent to others. The mining license shall be renewed, altered, and canceled according to law.

Security Acceptance Letter is a necessary certificate for mining enterprises, construction enterprises, and enterprises producing hazardous chemicals, fireworks and firecrackers, and civil explosives.

1. The basic licenses of the mine are *Business License*, *Mining License,* and *Security Acceptance Letter.*

    (1) *Business License* is legal and valid, and the enterprises' names and legal representatives are consistent with reality.

    (2) *Mining License* is legal and valid, and there is no phenomenon of super-layer transboundary and super-scale transboundary. The mining types, mining methods, production scale, and validity period shall be checked.

Whether the mine is over-exploited: The evaluation comments and conclusions of *Mine Reserves Dynamic Inspection Report* conclude that "there are no illegal mining activities beyond the boundary after an on-site inspection."

It is also possible to trust the mining enterprises to provide the special certificate of "whether it is over-exploited" and the original special certificate (copies stamped with official seal) from the authority of the natural resource of the county (city or district) to which they belong.

Mining methods: Check the "development and utilization program" and assessment opinions.

Mining the main minerals: Check *Resource Reserve Report*, *Development and Utilization Program* and Evaluation Opinion, and the on-site review, which shall be strictly consistent.

The development and utilization program of mining resources production is consistent with the preliminary design program or project construction. For example, "the mine adopts open-pit mining; the scale is 600,000 t/a; set one at intervals of one (or two), with the top-down stepped mining with the step's height of 15 m and below, the safety platform's width of 4 m, and the cleaning platform's width of 8 m (or 6 m); "road development and automobile transportation" shall be consistent in resource development and utilization program and preliminary design program or project construction.

(3) During the Security Acceptance Letter's validity period, the enterprises' names and legal representatives are consistent with the mining license. *Security Acceptance Letter* is generally issued by the production system. When there are multiple mining areas, some have safety production license, and some can also have construction period certificate. *Security Acceptance Letter* requires the system size to be consistent.

Mines that fail to issue *Security Acceptance Letter* by the safety production supervision and management department, such as underground water drawings, may not verify *Security Acceptance Letter*.

2. In addition to the above certificates, mining enterprises shall also apply for *Sewage Discharge License*, *Water Drawing License*, and *Civilian Blasting License*.

Within the validity period of the *Sewage Discharge License*, the pollutants discharged by the enterprise shall be consistent with the requirements of the *Sewage Discharge License*.

During the Water Drawing License's validity period, the water source and water amount shall be consistent with the *Water Drawing License*.

Civilian Blasting License, *Miner Qualification Certificate*, *Tree Logging License*, and other licenses are within the validity period.

Environmental protection facilities shall be subject to administrative examination and approval and shall be checked and accepted for compliance following relevant regulations, involving changes that have completed corresponding procedures.

Onsite evaluation or supporting materials:

For the mining enterprises' original business license, mining license, and safety production license, the enterprises shall provide the photocopy stamped with the official seal as supporting materials.

II. No administrative penalty within 3 years

Requirements: No administrative penalties have been imposed by natural resources and ecological environment departments in the past 3 years (3 years prior to the date of issuance of the selection notice), or the publishments have been rectified (certified by relevant management departments), and no significant safety or environmental protection accidents have occurred.

[Interpretation] This article stipulates that mining enterprises must operate following the law.

This article's requirements are the rejection requirements of the evaluation indicators determined according to the requirements of "4.1 Mining Enterprises shall abide by national laws and regulations and relevant industrial policies, and conduct mining following the law" in the *Green Mine Construction Specification.*

1. Relevant regulations on administrative penalties

   Major administrative penalties such as confiscation of illegal income, confiscation of illegal property, order to suspend production and business, temporary detention or revocation of permits, temporary detention or revocation of licenses, and administrative detention are relatively administrative severe penalties. Please refer to the Administrative Penalty Law of the People's Republic of China and Administrative Punishment Measures for Land and Resources for relevant regulations.

   (1) On September 1, 2017, at the 29th session of the Standing Committee of the Twelfth National People's Congress, No. 76 Order of The Presidential of the People's Republic of China issued *Decision of the NPC Standing Committee on Amending "Judge Law of the People's Republic of China" and other eight laws*, which came into force from January 1, 2018. Among them, the *Administrative Punishment Law of the People's Republic of China* has been amended for the second time, and the amendments related to this requirement are as follows.

       Chapter 2 Types and Settings of Administrative Penalties. Article 8 Types of administrative penalties:

       (1) Warnings;
       (2) Fines;
       (3) Confiscate illegal gains and illegal properties;

    (4) Order to suspend production and business;
    (5) Temporarily suspend or revoke licenses and permits;
    (6) Administrative detention;
    (7) Other administrative penalties prescribed by laws and administrative regulations.

(2) Administrative Punishment Measures for Land and Resources came into effect on July 1, 2014.
Article 4 Administrative penalties for land and resources include:
    (1) Warnings;
    (2) Fines;
    (3) Confiscate illegal gains and illegal properties;
    (4) Dismantle within a time limit;
    (5) Revoke exploration permits and mining licenses;
    (6) Other administrative penalties prescribed by laws and regulations.

2. "Within the last 3 years" refers to the 3 years before the issuance of the selection notice, that is, 36 months before the issuance of the selection notice.
3. Requirements relating to penalties.

    (1) In the past 3 years, it has not been punished by the natural resources and ecological environment departments.
    (2) It has been punished by the natural resources and ecological environment departments in the past 3 years but has been rectified.
    The relevant department shall issue a certificate of no penalty or a rectification certificate issued by the penalized enterprises. In particular, enterprises punished many times shall have proof of multiple rectifications in place. The national green mine hotline service platform of Zhongguancun Green Mine Industry Alliance can provide all the penalty data from the ecological and environmental protection department of mining enterprises for natural resource departments at all levels for free. The assessed enterprises that have been punished multiple times within 3 years shall be clearly identified, and detailed descriptions shall be made in the evaluation documents for the reference and decision-making of relevant departments.
    (3) If punished by the natural resources and ecological environment departments during the selection period, proof of rectification is needed.

4. Major accidents

    (1) Major safety accidents.
    According to Article 3 of *Regulations on Reporting, Investigation, and Handling of Production Safety Accidents*, a major accident refers to an accident that causes the death of more than ten people and less than 30 people, or severe injuries of more than 50 people and less than 100 people, or accidents involving direct economic losses of more than 50 million yuan and less than 100 million yuan.
    (2) Major environmental pollution accidents.

Major environmental pollution accidents refer to the contamination accidents that result in the death of more than ten people and less than 30 people, or the poisoning (serious injuries) of more than 50 people and less than 100 people. Where the regional ecological function is partially lost or the living environment of endangered species is polluted; where the local economic and social activities have been greatly affected due to environmental pollution, and the evacuation of people is from 10,000 to 50,000; where type-1 and -2 radioactive sources are lost, stolen or out of control; where large areas of major rivers, lakes, and reservoirs are polluted due to environmental pollution, or where water supply from water sources in cities and towns above the county level is interrupted.

5. Other related requirements

   (1) Two mining rights under an enterprise. If one mining right has been administratively punished for some reason, the green mine declaration of the other mineral right will not be affected by this.
   (2) If an enterprise related unit has received an administrative punishment, its declaration of green mine will not be affected.
   (3) If an enterprise legal person or shareholder has received an administrative penalty, the enterprise's declaration of a green mine will not be affected.

Onsite evaluation or supporting materials:

(1) The mining enterprise shall provide non-punishment certificates issued by the county-level natural resources and ecological environment department or rectification completed certificates issued by the penalty unit.
(2) Check the original of the special certificate, and take the photocopy with the official seal as the supporting materials.
(3) Query related information online.

III. Directory of abnormal mining owners

Requirements: During the selection period, the mining owners shall publicize the mining owners' exploration and mining information, and the mining owners shall not be included in the abnormal directory of the mining owners' exploration and mining information disclosure system.

[Interpretation] This article stipulates abnormal mining owners' directory not included during the green mine application or selection period.

This article's requirements are the rejection requirements of the evaluation indicators determined according to the requirements of "4.1 Mining Enterprises shall abide by national laws and regulations and relevant industrial policies, and conduct mining following the law" in the *Green Mine Construction Specification.*

When the mines participate in the green-mine application or selection period, the mining owners are not included in the abnormal mining owners' directory.

1. Abnormal directory
   Enter the national mining owners' exploration and mining information publicity system (http://kyqgs.mnr.gov.cn/) and input the mining owners' names or the unified credit code to check the mining owners' abnormal situation.

2. Conditions for listing and removing the abnormal directory.
   Article 23 of the Measures for Publicity of Information on Exploration and Mining by Mining Owners (Trial) stipulates that: if mining owners are under any of the following circumstances, the competent department of land and resources shall decide to include it in the abnormal directory within 20 working days, and make an announcement through the information publicity system to remind them to fulfill relevant obligations.

   (1) The mining owners fail to publicize the annual information following the provisions of the measures as of the end of filling annual information.
   (2) The competent department of land and resources discovers and verifies mining owners concealing or falsifying annual information.
   (3) The competent department of land and resources discovers and verifies that the mining owners have not fulfilled legal obligations or have not fulfilled legal obligations in place.

   Article 24 stipulates that: Within 3 years from the date when the mining owners are included in the abnormal list if the following circumstances are met, the land and resources department that decides to include it shall remove it from the abnormal directory.

   (1) Anyone is included in the abnormal directory because "Mining owners failed to publish annual information according to the provisions of these measures," and has made the supplementary declaration of unreported annual information and announced.
   (2) Anyone has been included in the abnormal directory because of "the annual information of mining owners conceal the true situation or falsify," and who has been corrected and announced.
   (3) Anyone has been included in the abnormal directory because "the mining owners have not fulfilled legal obligations or have not fulfilled legal obligations in place" and have performed obligations according to regulations and made public announcements.

3. The relevant provisions of *Provisions on Environmental Protection of Mines* (following the third amendment of Decision of the Ministry of Natural Resources on the First Batch of Repealed and Amended Departmental Regulations).
   Article 26 Anyone who violates these regulations and fails to compile the mining geological, environmental protection, and land reclamation plan, or expands the mining scale and alters the mining area's scope or the mining method, or fails to

reorganize the mine geological, environmental protection, and land reclamation plan and be approved by the original approval authority, shall be ordered for corrections within a time limit and be included in the abnormal directory or list of serious violations. Those who fail to correct within the time limit shall be fined less than 30,000 yuan and shall not be accepted to apply for new mining rights licenses or applications for mining licenses' renewal, alteration, and cancellation.

Article 27 Anyone who violates these regulations and fails to follow the approved mining geological, environmental protection, and land reclamation plan, or fails to complete the treatment and restoration before the mine approved to shut or close, shall be ordered for corrections within a time limit and be included in the abnormal directory or list of serious violations. Those who refuse to correct within the time limit or the corrections are not in place shall be fined less than 30,000 yuan and shall not be accepted to apply for new mining rights licenses or applications for mining licenses' renewal, alteration, and cancellation.

Onsite evaluation or supporting materials:

(1) Mining enterprises provide screenshots for the online query of abnormal situations of mining owners.
(2) Onsite network query screenshot materials (the dates should be noted).

IV. Mines' requirements

Requirements: The mines shall be in normal operation, and the mining time limits of remaining reserves (according to the annual report of reserves) shall be not less than 3 years.

[Interpretation] This article involves the requirements for the normal operating hours and resource reserves of enterprises declaring green mines.

This requirement contains two aspects, namely:

"Normal operation of the mine" refers to the mine's construction project's completion and acceptance with complete licenses and normal production. Firstly, the project is required to be completed and accepted with complete licenses. Then there is no long-term suspension of production or in resuming production.

"Mining time limits of remaining reserves" refers to the remaining mining time limits approved by the reserve report and production scale in principle.

Onsite evaluation and supporting materials:

Annual report of reserves, reserve resources' certificate issued by the natural resources department.

V. Scope of mining areas

Requirements: The scope and location of mining areas shall not involve natural reserves.

[Interpretation] This article involves the requirements for the mining areas' scope and location.

Guidelines on Establishing Natural Reserve System Dominated by National Parks issued and implemented by the General Office of the CPC Central Committee and General Office of the State Council on June 2019 are guiding opinions for accelerating the establishment of a natural protection system dominated by national parks, providing high-quality ecological products, and promoting the construction of a beautiful China.

National park refers to the specific land or sea area to protect the representative natural ecosystem and realize the scientific protection and rational utilization of natural resources. It covers the most important, the unique natural landscape and the most quintessence of natural heritage parts of China's natural ecosystem with an enormous scope of the protected area, complete ecological process, global value, national symbol, and a high degree of national recognition.

Nature reserve refers to areas that protect distinct natural ecosystems, areas where rare and endangered wild animal and plant species are naturally concentrated, and areas with the natural heritage of special significance. It has a large area to ensure the safety of the main protection objects, and maintains and restores the populations of rare and endangered wild animals and plants and the habitat that they depend on for survival.

Natural park refers to areas that protect important natural ecosystems, natural relics, and natural landscapes for sustainable use, with ecological, ornamental, cultural, and scientific value. Ensure the precious natural resources (such as forests, oceans, wetlands, waters, glaciers, grasslands, and living things), landscapes, geology, landforms, and cultural diversity be effectively protected. It includes natural parks such as forest parks, geological parks, marine parks, and wetland parks.

Proof materials to be provided:

Mining enterprises shall provide the special certificates issued by the competent authorities, such as the Natural Resources Bureau of the county (city or district), and take photocopies with an official seal as the proof materials.

Chapter 3
Green Mine Construction Rating Sheet

J. Deng et al., *Interpretation of Green Mine Evaluation Index*,
https://doi.org/10.1007/978-981-16-5433-6_3

First level	Second level	Level three indicators	Standard score	Grading description	Assessment method	Basis or standard	Inspection record	Score
1. Mine environment	Mine appearance	1. Function division	10	① The site is divided into functional areas according to the production area, management area, and living area, and 5 points are awarded for meeting zoning requirements. ② Five points are awarded for keeping a certain distance from the living area, such as the gangue dump, waste dump, waste yard, slag dump site, and mineral processing field	Search information, and check the site	General plan or diagram of the mining area		
		2. Production supporting facilities	15	Ground transportation, water supply, power supply, and other supporting facilities in the mining area shall be complete and in normal operation. Five points are deducted for any incomplete equipment or functions	Search information, and check the site	The general plan of the mining area		
		3. Living facilities	15	Staff dormitories, canteens, bathhouses, toilets, and other facilities shall be fully equipped, clean and tidy, and well managed. Five points are deducted for any not up to standard	Check the site			
		4. Production area signs	15	① The production area shall set up signs such as operation reminder signs, instruction signs, circuit diagram signs as required. 3 points are deducted for any spot not marked ② The size, shape, and color of the signs shall be set following the regulations, and 3 points are deducted for any unqualified	Check the site	*Signs* (GB 13306), and *Mine Safety Sgns* (GB 14161)		
		5. Customized management	15	Equipment and materials are managed in a standardized manner to achieve classification and partitioning, orderly arrangement, and neat stacking, and 5 points are deducted if any equipment and materials are littered or managed in disorder	Check the site			
		6. Solid waste stacking	7	① 3 points are awarded for solid waste with fixed storage sites ② Four points are awarded for solid waste with normative storage sites	Check the site	*General Industrial Solid Waste Storage and Disposal Sites Pollution Control Standard* (GB 18599), and *Pollution Control Standard For Hazardous Waste Storage* (GB 18597)		

(continued)

(continued)

First level	Second level	Level three indicators	Standard score	Grading description	Assessment method	Basis or standard	Inspection record	Score
		7. solid waste management	8	Solid waste storage sites shall be in normative operation and management with pollution control in place, and 8 points are awarded for no seepage and no domestic garbage mixed in	Check the site			
		8. Disposal and utilization of domestic waste	20	① Five points are awarded for domestic garbage in the mining area (including mines) collected in a fixed location ② Five points are awarded for classifying domestic garbage and reasonably determining the garbage's classification scopes, varieties, requirements, collection and transportation methods ③ Ten points are awarded for harmless disposal of domestic waste by oneself or by entrusting a third party, and providing proof materials	Check the site			
		9. Main road conditions	15	The main road surface in the mining area meets the specifications, and the surface is smooth, dense, and appropriate in roughness. Eight points are awarded for meeting the specifications, and Seven points are awarded for good maintenance	Check the site	*Code for design of roads in factories and mines* (GBJ22)		
		10. Cleaning condition of roads	10	The internal roads or special roads in the mining area have no spillage; or take effective measures to clean up the spillage in time. Five points are deducted for any unqualified location	Check the site			
		11. Cleaning condition of mining areas	20	The mining area shall keep clean and sanitary and there is no garbage or waste rock thrown randomly in the production area and management area, and there is no running, leaking, dripping or leaking of the pipelines on the production site. 5 points are deducted for any unqualified location	Check the site			
		12. Construction and maintenance of mining buildings and structures	20	① Twelve points are awarded for no temporary buildings or abandoned structures such as unauthorized construction in all places in the production area, management area, and living area; four points are deducted for any unqualified location found ② Eight points for timely maintenance, repair, or painting of buildings and structures in mining areas. Two points are deducted for any obvious damage, aging, and no repair or maintenance measures are taken	Check the site			
	Mine greening	13. Green coverage in the mining area	20	The mining area's green area shall be fully covered, and no exposed surface soil of the large area. Five points are deducted for any spot that does not meet the requirements	Check the site			

(continued)

(continued)

First level	Second level	Level three indicators	Standard score	Grading description	Assessment method	Basis or standard	Inspection record	Score
		14. Special main road greening and beautification requirements	10	Ten points are awarded for setting landscaping on both sides of the entry road, the road inside the office area, and the road from *the office area to the mining* area's production area as follows ① Isolated green belts shall be set up if the conditions are met, and greening shall be carried out according to local conditions ② Beautify the environment and make publicity boards or slogans if there are objectively no greening conditions	Check the site			
		15. Greening guarantee mechanism	4	Mine greening shall have a long-term guarantee mechanism, a greening maintenance plan and a responsible person, and four points are awarded for meeting the requirements	Search information, and check the site			
		16. Protection effect of greening	6	Greening plants collocation shall be reasonable, and six points are awarded for no serious dead branches and yellow leaves and no missing or dead seedlings	Check the site			
		17. Mine beautifications	10	Ten points are awarded for building parks, gardens, green spaces, and other landscape facilities by making full use of the mining area's natural conditions and topography and landforms according to local conditions	Check the site			

(continued)

(continued)

First level	Second level	Level three indicators	Standard score	Grading description	Assessment method	Basis or standard	Inspection record	Score
2. Resource development methods	Resource exploitation	18. Mining technology	50	★ Suitable for open mining: ① Drilling: Use wet drilling, dry drilling (with dust collection), and other rock drilling operations to drill ② Blasting: Use millisecond blasting, pre-split blasting, and smooth blasting methods ③ Shovel loading: Use large-scale automated hydraulic shovel loading equipment, hydraulic excavators or loaders, dump trucks, large-scale self-moving crushers, and other advanced equipment for shovel loading operations ④ Dumping: During the production period, use the technology of internal drainage by stages to maximize the use of internal drainage and reduce the external land occupation Fifty points are awarded for meeting all the requirements, and those that are not involved are deemed to meet the requirements; 20 points are deducted for any undesirable until 50 points are deducted (If both underground and open-pit mining are performed, choose one of the main mining methods at the current stage for scoring, and the scores shall not be accumulated)	Search information, and check the site			
				★ Suitable for underground mining: ① Use the filling method and water conservation mining techniques for underground mining ② It can effectively reduce the large-scale ground subsidence caused by mining ③Use the goaf to dispose tailings, waste rock, coal gangue on a large scale Fifty points are awarded for meeting all the requirements, and those not involved shall be deemed to meet all the requirements; 20 points are deducted for any undesirable until 50 points are deducted (If both underground and open-pit mining are performed, choose one of the main mining methods at the current stage for scoring, and the scores shall not be accumulated)	Search information, and check the site			

(continued)

(continued)

First level	Second level	Level three indicators	Standard score	Grading description	Assessment method	Basis or standard	Inspection record	Score
				★ Suitable for petroleum, natural gas, geothermal water, and other minerals: ① Use electric drilling machine and top drive device ② Use optimized and fast, controlled pressure and other drilling technologies ③ Use environmental friendly drilling fluid and recycling technology Dispose drilling mud and other drilling wastes timely and harmlessly Fifteen points are deducted for any undesirable until 50 points are deducted	Search information, and check the site			
		19. Quality requirements of mining face	30	★ Suitable for open mining: ① The operating platform shall be clean, smooth and unobstructed, free of debris and water, and free of dangerous rocks on the slopes of the working steps and non-working steps; 15 points for meeting the requirements ② Clean up the falling objects on the non-working steps in time, and set a warning line or safety sign at safety hazard position; 15 points for meeting the requirements	Check the site			
				★ Suitable for underground mining: ① The safety exit of underground mining face is unimpeded to meet the needs of ventilation, transportation, pedestrians, equipment installation and maintenance, and the support is intact. Fifteen points are awarded for meeting the requirements ② There is no large area of water, no floating ballast, and no debris on the working face, and the materials are neatly stacked. Fifteen points are awarded for meeting the requirements	Check the site			

(continued)

(continued)

First level	Second level	Level three indicators	Standard score	Grading description	Assessment method	Basis or standard	Inspection record	Score
				★ Suitable for petroleum gas and mineral water: ① There is no leakage or throwing of hazardous chemicals to prevent "running and dripping" and pollution of the well site's surface soil ② Drilling waste shall not fall to the ground, and carry out centralized and harmless treatment ③ Regularly spray water on the well site's bare ground for dust suppression Ten points are awarded for each item meeting the requirements	Check the site			
	Mineral dressing and processing	20. Mineral dressing and processing technology	60	★ Suitable for nonferrous metals, metallurgy, gold, nonmetal, chemical, coal, and other industries: ① Use production equipment and processes with a high degree of automation, low energy consumption, and low pollutant production				
				② The beneficiation indicators such as the beneficiation recovery rate, concentrate grade, and the grade meet or exceed the design requirements, and the main metals and associated elements are fully utilized ③ Use high-efficiency and low-toxic agents with little impact on the environment (for example, the cyanide chemical chamber in the gold industry shall be isolated and completely closed) ④ The content of valuable components in tailings and waste rock shall not be higher than the grade that can be processed by the current technical level Fifteen points are deducted for any undesirable until 60 points are deducted	Search information, and check the site			
				★ Suitable for cement limestone industry: ① The production flow reflects a short process, low energy consumption, and high efficiency ② The crushing system selects the suitable, efficient crusher according to the rock's breakability ③ Crushing workshop, transportation corridor, and other main production areas shall be completely closed and equipped with dust collecting and dust removal equipment Twenty points are deducted for any undesirable	Search information, and check the site			

(continued)

(continued)

First level	Second level	Level three indicators	Standard score	Grading description	Assessment method	Basis or standard	Inspection record	Score
				★ Suitable for sandstone and building stone industry: ① Select short process, low energy consumption processes and equipment according to parent rock material properties, product structure, capacity requirements, and other factors, and configure auxiliary facilities following production scale and process ② Dry production is equipped with dust removal equipment and keeps running synchronously with production equipment, and wet production is equipped with mud powder and water separation, wastewater treatment and recycling systems ③ The dust-producing point in the production area shall be closed ④ Stone and gravel aggregate finished product storage yard (warehouse) has hardened ground, with storage classified or stored in separate warehouses Fifteen points are deducted for any undesirable	Search information, and check the site	*Design Specification for Mechanism Sandstone Aggregate Factory* (GB 51186)		
				★ Suitable for oil and gas, geothermal, mineral water, and other industries: ① Thirty points are awarded for selecting reasonable crude oil dehydration technology and equipment for dehydration, and select reasonable oil and gas separation equipment and crude oil stabilization technology ② Thirty points are awarded for taking effective measures for oil and gas reservoirs associated with carbon dioxide gas and hydrogen sulfide gas, and the associated gas's content not meeting the requirements for comprehensive industrial utilization	Search information, and check the site			
	Mine environment restoration and land reclamation	21. Scope requirements	30	The prescribed areas shall be treated and reclaimed according to the mine geological environment restoration treatment and land reclamation plan, such as dumps, open pits, special mining roads, industrial mining sites, subsidence areas, gangue dumps, and the contaminated sites of mines Five points are deducted for areas that shall be treated and reclaimed but fail to do in time according to the plan	Search information, and check the site	*Mine Geological Environment Protection and Land Reclamation Plan*		

(continued)

(continued)

First level	Second level	Level three indicators	Standard score	Grading description	Assessment method	Basis or standard	Inspection record	Score
		22. Governance requirements	10	① Sites after restoration shall be coordinated with the surrounding natural environment and have a landscape effect ② If the exposed area caused by open-pit mining has a significant impact on the surrounding landscapes, measures shall be taken to mitigate the adverse impact ③ The open-pit mining shall also meet the requirements of retaining platform and reclamation greening after open-pit mining Four points are deducted if one of the above three findings fails to meet the requirements until ten points are deducted	Search information, and check the site	*Mine Geological Environment Protection and Land Reclamation Plan, Land Reclamation Quality Control Standards* (TDT1036), and other supporting documents		
		23. Land use function requirements	10	Sites after the treatment shall recover the basic land function and realize the sustainable land use according to local conditions, and 10 points are awarded for meeting the requirements	Search information, and check the site	*Mine Geological Environment Protection and Land Reclamation Plan, Land Reclamation Quality Control Standards*(TDT1036)and other supporting documents		
		24. Ecological functional requirements	10	Sites after treatment shall meet: ① The overall ecological function of the region shall be protected and restored ② No threat to animals or plants. Five points are deducted for any undesirable	Search information, and check the site	*Mine Geological Environment Protection and Land Reclamation Plan, Land Reclamation Quality Control Standards*(TDT1036)and other supporting documents		

(continued)

(continued)

First level	Second level	Level three indicators	Standard score	Grading description	Assessment method	Basis or standard	Inspection record	Score
	Environmental management and monitoring	25. Environmental protection facilities	6	① Four points are awarded for complete environmental protection facilities and effective operation of relevant facilities ② Two points are awarded for effective maintenance	Search information, and check the site	Acceptance data of environmental protection facilities		
		26. Environment management system certification	4	Four points are awarded for obtaining environment management system certification	Check the certificates	ISO environmental management system certification		
		27. Environmental monitoring system	5	Establishing a long-term mechanism, and five points are awarded for having an environmental monitoring system	Search information	Environmental monitoring system		
		28. environment monitoring equipment	5	Five points are awarded for setting up automatic monitoring and electronic display equipment for noise and air pollutants	Check the site			
		29. Emergency response mechanism	5	Construct an emergency response mechanism, and five points are awarded for having emergency response measures responding to environmental emergencies	Search information	Emergency response institution		
		30. Dynamic monitoring of mine's geological environment	5	Five points are awarded for dynamic monitoring of mine's geological environment, such as ground deformation	Search information, and check the site	Dynamic monitoring record		
		31. Dynamic monitoring of wastewater and tailings.	5	5 points are awarded for dynamic monitoring of beneficiation wastewater, mine water, tailings (gangue hills), dumps, waste rock dumps, dust, and noises [sounds]	Search information, and check the site	Dynamic monitoring record		
		32. Dynamic Monitoring of Reclamation Area	5	Five points are awarded for dynamic monitoring of land damage, stability, soil quality, and reclamation quality in the reclamation area	Search information, and check the site	Dynamic monitoring record		
3. Comprehensive utilization of resources	(1) Non-metal, chemical, gold, metallurgy, non-ferrous, petroleum, coal and other industries are scored according to 10 three-level indicators of 33-42 items, with a total score of 120							

(continued)

(continued)

First level	Second level	Level three indicators	Standard score	Grading description	Assessment method	Basis or standard	Inspection record	Score
	Comprehensive utilization of symbiotic and concomitant resources	33. Resource exploration and evaluation and development	10	Ten points are awarded for comprehensive exploration, comprehensive evaluation, and comprehensive development of symbiotic and concomitant resources according to the plan for mineral resources' exploitation and utilization	Search information	*Mineral Resources' Development and Utilization Plan* and relevant product information		
		34. Comprehensive utilization of symbiotic and concomitant resources	20	Use advanced, applicable, economical and reasonable technology to process and comprehensively utilize the symbiotic and concomitant resources, and 20 points are awarded for meeting the requirements	Search information, and check the site	Production statements or financial statements		
		35. Comprehensive utilization of complex or low-grade ores	5	Use new technology to reduce energy consumption, or adopt a combined process of beneficiation and smelting to improve technical and economic indicators, and five points are awarded for obtaining results and providing proof materials	Search information, and check the site			
		36. Requirements for symbiotic and concomitant resources that cannot be exploited and utilized for the time being	5	Five points are awarded for taking effective protective measures for symbiotic and concomitant resources that cannot be exploited and utilized for the time being	Search information	*Mineral Resources' Development and Utilization Plan*		
	Solid waste disposal and comprehensive utilization	37. Industrial solid waste disposal and utilization	25	Establish comprehensive utilization of solid waste such as waste rock (slag), coal gangue, tailings, drilling waste mud, rock cuttings, scum, oil sludge, and 25 points are awarded for making full use of solid waste through backfilling, paving, and producing building materials	Search information, and check the site	*Mineral resources' development and utilization plan* and other proof materials		
		38. Disposal and utilization of topsoil	10	Ten points are awarded for stripping topsoil and overlying rocks for land reclamation and ecological restoration (No topsoil and overlying rocks shall not be scored, and 35 points are awarded for "37 Industrial Solid Waste Disposal and Utilization")	Search information, and check the site	*Mineral Resources' Development and Utilization Plan* and other proof materials		
		39. Recover and extract valuable elements/useful minerals	5	Five points for extracting valuable elements or useful minerals from solid wastes such as tailings, coal gangue, and waste rock	Check the site	Production statement, sales statement, and financial statement		

(continued)

(continued)

First level	Second level	Level three indicators	Standard score	Grading description	Assessment method	Basis or standard	Inspection record	Score
	Wastewater treatment and comprehensive utilization	40. Treatment and comprehensive utilization of mining wastewater	15	① Seven points are awarded for equipping with mining wastewater treatment facilities such as mine water, drain water, drilling wastewater, and well-washing wastewater ② 8 points are awarded for realizing the effective disposal of waste by adopting clean and resource technologies	Search information, and check the site	Production statement (scheduling report) or other proof materials		
		41. Treatment and comprehensive utilization of production wastewater	15	① Seven points are awarded for establishing a recycling treatment system for production wastewater such as mineral processing wastewater ② Eight points are awarded for recycling production wastewater	Search information, and check the site	Production statement (scheduling report) or other proof materials		
		42. Disposal of domestic sewage	10	① Four points are awarded for equipping with domestic sewage treatment system ② Six points are awarded for the effective disposal of domestic sewage	Search information, and check the site	Production statement (scheduling report) or other proof materials		
	(2) The sandstone, cement limestone and building stone industries are scored according to 4 three-level indicators of 43-46 items, with a total score of 120 points							
	Comprehensive utilization	43. Comprehensive utilization of mining and processing products	40	★ Suitable for aggregate and building stone industry: Forty points are awarded for making full use of stone powder, mud powder and other mining or processing products to improve the resource utilization level, such as new construction materials, engineering materials, environmental treatment and land reclamation, and soil improvement	Search information, and check the site	Production statement (scheduling report) or other proof materials		
				★ Suitable for cement limestone industry: Through combining the characteristics of the various raw materials and ingredients of the cement production line, realize the resource utilization of various products from mining or processing [production], the classified utilization of resources, high-quality and optimal use, and high-grade and low-grade ore, interlayer, roof and floor surrounding rocks. Forty points are rewarded for comprehensive utilization	Search information, and check the site	Production statement (scheduling report) or other proof materials		

(continued)

(continued)

First level	Second level	Level three indicators	Standard score	Grading description	Assessment method	Basis or standard	Inspection record	Score
	Solid waste disposal and comprehensive utilization	44. Comprehensive utilization of soil stripping	40	★ Suitable for aggregate and building stone industry: Forty points are awarded for the stripped topsoil or the sieved ballast soil and waste rock stacked in the dumpsite to produce new building materials, environmental management, land reclamation, ecological restoration, and other resource utilization methods	Search information, and check the site	Production statement (scheduling report) or other proof materials		
				★ Suitable for cement limestone industry: Twenty points are awarded for using qualified soil stripping material for silica-alumina raw materials or reclamation, and 20 points are awarded for using other stripping materials as cement ingredients, sand and gravel aggregates, or other engineering materials	Search information, and check the site	Production statement (scheduling report) or other proof materials		
	Wastewater treatment and comprehensive utilization	45. Treatment and utilization of production wastewater	30	① Ten points are awarded for equipping with a complete production wastewater treatment system ② The wastewater shall be treated to solid-liquid separation, and 20 points are awarded for efficient recycling of clean water	Check the site	Production statement (scheduling report) or other proof materials		
		46. Disposal of domestic sewage	10	① Four points are awarded for equipping with a domestic sewage treatment system ② Six points are awarded for the efficient disposal of domestic sewage	Search information, and check the site	Production statement (scheduling report) or other proof materials		
4. Energy conservation and emission reduction	Energy conservation and consumption reduction	47. Whole-process energy-consumption accounting system	5	Five points for establishing a whole-process energy consumption management system	Search information	Documents or ledger of the whole-process energy-consumption accounting system		
		48. Energy management plans	10	① Five points are awarded for an annual energy management plan ② Five points are awarded for deposing energy-saving indicators to subordinate units, departments, or workshops	Search information	Energy analysis report		

(continued)

(continued)

First level	Second level	Level three indicators	Standard score	Grading description	Assessment method	Basis or standard	Inspection record	Score
		49. Mine energy consumption per unit product	15	Five points are deducted for each item if energy consumption, material consumption, and water consumption per unit product fails to meet the specified requirements If there are national standards for coal mines, iron ore mines, gold mines, and non-ferrous metal mines, the national standards shall apply. For other minerals without national standards or industry standards for the time being, the assessment shall be based on the average value of the enterprises' energy consumption and other indicators in the past three years, and the requirements for energy saving and consumption reduction shall be reflected	Search information	Energy consumption ledger and energy consumption quota of each industry unit product		
		50. Energy management system certifications	5	Five points are awarded for enterprises obtaining the energy management system certifications	Check the certificates	Energy management system certificates		
	Exhaust emission	51. List of major dust-producing sites	5	The mine shall have an exact list of dust concentration in the main dust-generating operation sites such as mining, transportation, and mineral processing (processing) and their positions	Check the site	Related measures of enterprise dust-prevention		
		52. Dust emission during production	15	① Reduce dust by adopting measures such as the integrated rock drilling and dust collection drilling rig or wet rock drilling technology during rock drilling ② Reduce dust by spraying water during blasting operations ② Add dust collecting equipment at fixed dust production point and maintain sufficient negative pressure to operate synchronously to suppress and deal with dust generated in the process of selection and processing Three points are deducted for any unqualified in rock drilling, blasting, rock (ore) breaking (crushing), screening, conveying, batching, and other key links or locations	Check the site and random check employee to know	Special dust removal scheme shall be provided for blasting, and this scheme shall not cover other loose rock open-pit coal mines.		
		53. Dust emission during surface transportation	15	Set up water spray or induction spray facilities along the transportation road or configure sprinklers to spray water regularly to reduce dust, ground vehicles and transportation equipment to take spray dust or sprinkling dust, and use sealed vehicles for outbound products to avoid dust along the road. Three points are deducted for any unqualified	Check the site			

(continued)

(continued)

First level	Second level	Level three indicators	Standard score	Grading description	Assessment method	Basis or standard	Inspection record	Score
		54. Dust emission of storage sites	10	① Five points are awarded for waste rock or ore turnover sites and storage sites equipping with supporting dustproof facilities ② Five points are awarded for achieving the anti-dust effect	Search information, and check the site	Enterprises' relevant dustproof measures		
		55. Other gases emission	10	Ten points are awarded for an exhaust gas purification system and standardized discharge for industrial waste gas containing other toxic and harmful substances (such as SO_2 and NO_x) in addition to dust generated during the mining and selection process	Search information	Monitoring report or test data		
	Wastewater discharge	56. Domestic sewage discharge	10	Ten points are awarded for domestic sewage discharged to the standard after treatment or discharged directly into the municipal sewage pipe network	Search information, and check the site	Acceptance data of sewage stations and other environmental protection facilities		
		57. Industrial wastewater discharge	15	Zero discharge of industrial wastewater is encouraged. Fifteen points are awarded if the water quality reaches the discharge standard after treatment	Search information, and check the site	Inspection data of environmental protection department		
		58. Drainage pipe setting	10	Ten points are awarded for cleaning pipelines laid separately and rainwater and sewage pipes set separately	Check the site			
		59. Discharge requirements for surface runoff water and leaching water	15	① The mine area construct rainwater interception (drainage) ditch and sedimentation tank, and the collected surface runoff water and leached water shall be discharged or treated for reuse after reaching the standard after precipitation, and 10 points will be awarded for meeting the requirements waste dump and gangue dump shall set intercepting (discharging) ditches, and five points are awarded for meeting the requirements	Check the site	Overall mining design		

(continued)

(continued)

First level	Second level	Level three indicators	Standard score	Grading description	Assessment method	Basis or standard	Inspection record	Score
	Solid waste emissions	60. Solid waste discharge requirements	30	For solid waste that cannot be used comprehensively ① Divide different categories of hazardous waste, general waste, and domestic waste, and ten points are awarded for classification ② Twenty points are awarded for disposing solid wastes by oneself or entrusting a qualified third party to dispose following national laws and standards	Search information, and check the site	*Law of the People's Republic of China on the Prevention and Control of Environmental Pollution by Solid Waste, General Industrial Solid Waste Storage and Disposal-Site Pollution Control Standards* (GB 18599), and *Hazardous-Waste Incineration, Storage, Landfill, and Pollution Control Standards* (GB 18484, 18597, and 18598)		
	Noise emission	61. List of major noise points	5	The mine has a list of major noise producing places and their positions, and onsite inspections can be carried out if necessary. Five points are awarded for meeting the requirements	Check the site			
		62. Noise disposal requirements	15	Performing noise reduction treatment for rock drilling, crushing, air pressure, and other high-noise equipment in the mining area; 15 points are awarded for equipping with noise reduction, vibration reduction, and vibration isolation measures	Check relevant monitoring reports			
		63. Noise emission requirements	10	Ten points are awarded for noise emission at the factory boundary up to standard	Check the site	*Environmental Noise Emission Standards for Industrial Enterprises* (GB 12348-2008)		

(continued)

(continued)

First level	Second level	Level three indicators	Standard score	Grading description	Assessment method	Basis or standard	Inspection record	Score
5. Scientific and Technological Innovation and Intelligent mines	Scientific and Technological Innovation	64. Technology R&D team	3	Three points are awarded for the technology R&D team of enterprises construction with full-time technical personnel	Search information	Science and technology management system		
		65. Technology research and development management system	3	Three points are awarded for technical research and development's rewards and management	Search information	Science and technology management system		
		66. Collaborative innovation system	6	Establish industry-university-research-application collaborative innovation system: ① Two points are awarded for establishing technological innovation cooperation relationships with scientific research institutes and universities, and sign cooperation agreements to establish enterprise technology platforms, including engineering technology centers, enterprise technology centers, key laboratories, academician expert workstations, and innovation studios ② Carry out technical research to support the development of the main business of the enterprises, and two points are awarded for having project document or project ledger material Improve the enterprise technological level, and two points are awarded for proof materials	Search information	Announcement documents of the competent authority, project approval documents, and project ledger		
		67. Science and Technology Awards	18	Eighteen points are awarded for enterprises' research projects or achievements receiving national awards; 12 points are awarded for provincial and ministerial awards; ten points are awarded for the *Social Science and Technology Awards Catalogue* of the National Awards Office. Awards should promote green mines' construction and reflect the name of the unit, and the total score shall not exceed 18 points	Search information	Announcement documents of the competent authority, project approval documents, and project ledger		
		68. R&D and technical renovation investment	6	The investment in R&D and technological transformation shall not be less than 1.5% of the main business income of the previous year. Six points are awarded for reaching 1.5%; five points for 1-1.5%; four points for 0.5-1%; two points for less than 0.5% and investing in technical innovation projects for employees	Search information	Check financial statements, subsidiary accounts, auxiliary accounts, or project ledger		

(continued)

(continued)

First level	Second level	Level three indicators	Standard score	Grading description	Assessment method	Basis or standard	Inspection record	Score
		69. Hi-tech enterprise qualification	3	Three points are awarded for obtaining the hi-tech enterprise certificates	Check the certificates	Obtaining the hi-tech enterprise certificates		
		70. Intellectual property rights	6	Two points are awarded for obtaining an invention patent within three years; one point is awarded for publishing a core journal paper; one point for a utility model or software copyright. All achievements shall reflect the unit's name, and the total score shall not exceed six points	Search information	Originals of patents and papers (or copies with official seals)		
		71. Advanced technology and equipment	20	Select mining and dressing technologies, techniques and equipment selected into *Catalog of Environmental Protection Technologies Encouraged by the State*, *Catalogue of Advanced and Applicable Technologies for Conservation and Comprehensive Utilization of Mineral Resources*, *Directory of National Advanced Pollution Prevention and Control Demonstration Technologies*, *Promotional catalog of advanced applicable technologies, processes, equipment and materials for safe production*, and *Catalogue of National Key Energy Conservation Technology Promotion* and *Recommended catalog of energy-saving electromechanical equipment (products)*. Ten points are awarded for each technique, craft, or equipment, and the total score shall not exceed 20 points	Search information, and check the site	Relevant industrial policy catalogs, design specifications, and relevant certification materials		
	Intelligent mines	72. Intelligent mines construction program	5	Two points are awarded for the content of intelligent mines construction in the enterprise's annual plan, and three points are awarded for carrying out as planned	Search information, and check the site	Enterprise's annual plan		
		73. Mine automation centralized control platform	10	Construct the mine automation centralized management and control platform, displaying the automatic control system, remote monitoring system, storage management system, and various monitoring systems centrally and uniformly; ten points are awarded for meeting the requirements	Check the site	The construction plan of mine automation centralized control system platform		

(continued)

(continued)

First level	Second level	Level three indicators	Standard score	Grading description	Assessment method	Basis or standard	Inspection record	Score
		74. Mine-production automation system	10	① Four points are awarded for establishing an unattended automation system for fixed facilities in places such as a central substation, water pump room, fan station, air compressor room, and belt transportation lane ②Three points are awarded for establishing a remote-control system for main equipment in the mining and production process Three points are awarded for establishing safety monitoring system platforms for waste rock yard, spoil area, slope construction, and working environment	Check the site	Mine automation subsystem construction scheme		
		75. Remote video monitor and control system	10	Establish a complete remote video monitoring system. The remote video monitoring system shall be installed in mine working face and other production sites, key points such as power supply, drainage, ventilation, transportation, measurement and sales, and important safety sites such as tailings ponds and roadways. One point is awarded for each installation and real-time monitoring, and the total score shall not exceed ten points	Search information, and check the site			
		76. Resource storage management system	5	Five points are awarded for carrying out the actual work of 3D storage management	Check the site			
		77. Intelligent working face or driverless mine car system	5	Five points are awarded for one of the following two items: ① Set up the intelligent working surface for normal production ② Construct driverless mine car system	Search information, and check the site	The design scheme of the intelligent working face or driverless mine car system		
		78. Mine environment on-line monitoring system	5	Establish an online environmental monitoring system in mining areas with the function of an automatic grading alarm and grading notification for emission indicators of pollutants (atmospheric pollutants from mine water and solid waste noise) under the supervision of the competent administrative department of environmental protection. Five points are awarded for meeting the requirements	Search information, and check the site	The construction scheme of mine environment on-line monitoring system		

(continued)

(continued)

First level	Second level	Level three indicators	Standard score	Grading description	Assessment method	Basis or standard	Inspection record	Score
6. Enterprise management and enterprise image	Green mine management system	79. Green mine construction plan and target	5	Five points are awarded for the enterprises' annual plan for green mine construction content, goals, indicators, and corresponding measures	Search information	Enterprises' annual plan		
		80. Organizational structure and responsibilities of green mine construction	5	Five points are awarded for clear organization and responsibility system of green mine construction	Search information	Relevant documents on the establishment and responsibilities of the green-mine management organization		
		81. Green mine assessment	5	Establish a green mine assessment mechanism and conduct internal assessments at least once a year, comparing to the green mine construction plan and goals. Five points are awarded for meeting the requirements	Search information			
		82. Improvement and promotion of green mine construction	5	Clarify the improvement content, measures, persons in charge, completion time, and effect achieved of green mine construction; five points are awarded for meeting the requirements	Search information			
		83. Green mine construction training	8	① One point is awarded for green mine training system and plan ② 3 points are awarded for organizing management and technical personnel to conduct training (learning) on green mine construction ③ Regularly organize full-time green mine personnel to participate in systematic training (learning) on green mine construction, and 4 points are awarded for Training (learning) certificate	Search information and random check employee to know	Training system, training plan, training sign-in, video materials, training notices, certificates, photos		
	Enterprise culture	84. Employee satisfaction survey	3	Carry out employee satisfaction questionnaire surveys regularly, set up reasonable questionnaire survey content to be objective and fair. One point is awarded for the organization once a year; one point for the satisfaction level higher than 70%; one point for timely publicity	Random check employee to know	Original records of questionnaires		

(continued)

(continued)

First level	Second level	Level three indicators	Standard score	Grading description	Assessment method	Basis or standard	Inspection record	Score
		85. Recreational activities of the staff	4	① Two points are awarded for staff leisure, entertainment, cultural, and sports facilities ② Two points are awarded for facilities operating normally	Search information, and check the site			
		86. Trade unions organize activities	3	Three points are awarded for the labor union regularly carrying out activities to promote exchanges between employees and enterprises	Search information			
		87. Green mining culture construction	3	Green mine promotional videos based on clarity, commentary, duration, and other key content, and give points according to the production effect	Watch promo-tional video			
	Enterprise management	88. The linkage mechanism between employee income and enterprise performance	2	Establish a mechanism for employees' income to increase synchronously with enterprise performance, and two points are awarded for the enterprise employees' total income related to the growth of enterprise economic efficiency	Search information and random check employee to know	Assessment System		
		89. Functional area management system	2	Two points are awarded for a production and living management system consistent with the enterprise's actual situation and clearly defined responsible unit or department	Search information	Check the related management documents of the mines		
		90. Selection and management of equipment	20	① There is a list of core equipment, including equipment name, model, main parameters, energy consumption, purchase time, and maintenance ② The onsite inspection equipment conforms to the list and shall be used normally, and there is no backward production technology equipment explicitly eliminated by the state Five points are awarded for meeting one item	Search information	Check the related management documents of the mines		
		91. Occupational health management system	3	Three points are awarded for equipping with occupational health and other management systems	Search information	Check the related management documents of the mines		

(continued)

(continued)

First level	Second level	Level three indicators	Standard score	Grading description	Assessment method	Basis or standard	Inspection record	Score
		92. Environmental-protection management system	3	Three points are awarded for equipping with environmental protection management system (including sewage and wastewater discharge; classification, stacking, and control of solid waste; noise control; dust control)	Search information	Check the related management documents of mines		
		93. Personnel visual management	4	① The internal employees shall wear uniform labor insurance clothing in compliance with safety requirements when entering the production site ② External personnel, such as visitors, inspectors, learning personnel, and contractor staff shall dress according to the safety requirements of the production site when entering the production site ③ One point is deducted for anyone or any place undesirable	Check the site	Personnel visual management system		
		94. Green mine publicity campaign	6	Carry out publicity activities related to green mines construction, publicity reports, publicity week activities, and contests, and publish positive reports in the media. Two points are awarded for each category, and the total score shall not exceed six points	Search information, and check the site			
		95. Employees' medical examination	4	Two points are awarded for enterprises organizing all employees to take regular physical examinations every year, and two points are awarded for classifying and formulating physical examination plans, physical examination items, and establishes occupational health surveillance files	Search information	Medical records		
	Community harmony	96. Mine's harmony	5	Establish a good relationship with the township (street), village (community) and handle all disputes and contradictions properly promptly	random check employee or visit the community			
		97. Poverty alleviation or charity fundraising activities	5	Enterprises carry out poverty alleviation or charity fundraising activities regularly or irregularly. Five points are awarded for carrying out poverty alleviation or charity fundraising activities in the past two years	Search information and random check employee to know	Poverty alleviation contract or donation contract or relevant proof of bills		

(continued)

(continued)

First level	Second level	Level three indicators	Standard score	Grading description	Assessment method	Basis or standard	Inspection record	Score
	Enterprises credit	98. The situation of tax payment by enterprises according to law	4	Enterprises shall pay taxes following the law, pay taxes in good faith, and pay taxes actively. Two points are deducted each time if there is tax evasion until 4 points are deducted	Investigation visit and search information	Tax department certificate		
		99. Enterprises' performance of relevant obligations	4	Enterprises shall collect and submit geological data as required; Submit the basic statistics table of mineral resources on time Two points are deducted for any undesirable	Search information			
		100. Information announcement	2	Two points are awarded for enterprises publicizing the exploration and mining information of mining owners according to the regulations	Information disclosure system for exploration and exploitation of mining owners			
Total points			1,000					

Chapter 4
Mining Area Environment

As a significant window to show the enterprise image to society, the mining area environmental construction is the internal need to create a healthy environment for workers' production and living and an essential part of the ecological system in the mining area. The layout, management, greening, and beautification of the mining area are the basic requirements of green mine construction, reflecting mining enterprises' social obligation.

The mining-area environment evaluation should reflect the mining area's social image and production and living security ability.

1 Mining Area Appearance

1 Functional zoning

Standard score: Ten

Instructions: (1) Five points are rewarded for properly dividing the site into the production zone, administrative zone, and living zone according to their functions. (2) Five points for keeping a safe distance between the waste rock dump, waste soil dump, waste dump, waste residue dump, ore dressing plant, and the living zone

Assessment method: Look up information, check the site

Criteria and standard: General layout or schematic diagram of the mining area

Explanations: This article requires the mining area's serviceability for living and production from space layout points.

J. Deng et al., *Interpretation of Green Mine Evaluation Index*,
https://doi.org/10.1007/978-981-16-5433-6_4

This evaluation requirement is formulated following "the functional zone dividing in the mining area should be reasonable, the mining area should be decorated with green vegetation, and the overall environment should be clean and beautiful" (See Sect. 5.1.1) in "Green mine construction principle" and "each functional zone shall comply with the requirements of orderly operation and standardized management in *Code for Design of General Plan of Industrial Enterprises* (GB 50,187–2012)" (See Sect. 5.2.1).

Item (1) will score five points for clear zoning, while there will be no points if the zone is divided unclearly or incompletely; item (2) will score no points if the stacking place of solid wastes is too close to the office and living area, leading to safety accidents or environmental pollution.

1. Reasons for functional zoning
   Functional zoning refers to various facilities' layout in industrial enterprises according to different functions and systems, forming an interconnected organic whole. From the production view, functional zoning is conducive to the centralized layout of production facilities to realize specialized production, improve production efficiency, and maximize the function of production system. From living, functional zoning is conducive to creating a safe, healthy and comfortable living environment, improving employees' life quality, and stimulating workers' enthusiasm.
2. According to the standard, the mining area should be divided into the production area, administrative area, living area, and ecological area.
   Functional zoning refers to various facilities' layout in industrial enterprises according to different functions and systems, forming an interconnected organic whole. According to their functions, the mining area is divided into the production zone, administrative zone, living zone, and ecological zone.
   Production zone refers to the area where the personnel, equipment, materials, and processing objects need organized control to meet the site's production process requirements, sanitation, and quality.
   Administrative zone: Administrative zone is the office operation and rest area where operators are organized to work according to the procedures, and staff's behaviors are standardized through monitoring, analysis, command, and communication. This area always includes office buildings, joint buildings, and a dispatching command center.
   Living zone is residential buildings and public cultural welfare facilities set up in the factory for employees' service, including a dormitory, family residence, canteen, kindergarten, bathroom, hospital, cinema, and club. The living zone refers to the staff dormitory and amenities in the mining area in a narrow sense. Generally, the living zone should be located on the windward side of the production zone.
   Ecological zone: Ecological zone refers to the area without project layouts or the area that has formed a stable ecosystem after land reclamation and will have no more project layouts in the mining area.
3. Functional zoning shall comply with the requirements of orderly operation and standardized management in *Code for Design of General Plan of Industrial Enterprises* (GB 50,187—2012).

The special evaluation on the storage and deposit site in EIA should comply with the regulation on general industrial solid wastes[1]—"solid waste sites should be located at the downwind side of the dominant wind in the industrial area and residential area, and the plant boundary[2] should be 500 m away from the residential area (See Sect. 5.1.2)", and "special evaluation of storage and disposal sites shall be arranged in construction project EIA; EIA procedures shall be re-performed for those expanded, rebuilt and extra-served storage and disposal sites (See Sect. 6.1.2)" in the *Standards for Pollution Control on the Storage and Disposal Sites for General Industrial Solid Wastes* (GB 18,599—2001).

According to relative requirements in *Code for Design of General Plan of Industrial Enterprises* (GB 50,187—2012) and onsite assessment, waste rock dump, waste soil dump, waste dump, and waste residue dump shall be separated from the office area and living area and reserve a safety distance.

4. The production zone, administrative zone, living zone, and solid-waste stacking place shall be marked on the general ground layout plan, remote sensing image map, aerial photo map, and schematic diagram of the mining area.

Onsite evaluation or supporting materials:

The general layout plan or other specially made real effect drawings can mark the production zone, administrative zone, living zone, and waste stacking place.

The mining area's general plan includes the production area, management area, living area, and waste piling location, or other real effect drawings specially made to mark the production area, management area, living area, and waste piling location.

2 Production facilities

Standard score: 15

Instructions: Supporting facilities such as ground transportation, water supply, and power supply shall be complete and function properly in the mining area. Five points are deducted for defective equipment

Assessment method: Look up information and check the site

Criteria or standard: The general layout plan

Explanations: This article is the requirement for production facilities **service capacity.**

1. Transportation infrastructure

According to the traffic flow, terrain and surface features, pedestrian overpass bridges (including underground crosswalks), fences, lighting facilities, sight

[1] Solid trash, solid waste material and solid waste are used in various laws, documents and standards. This book uses solid waste in the non-quoted part.

[2] Plant boundary refers to the place or building boundary with the use (or ownership) right determined by legal documents (such as land use certificate, house property certificate, and lease contract).

guidance signs, emergency contact facilities, and other similar facilities shall be set on the road when necessary.

2. Power supply facilities
   Power supply facilities, also known as electric facilities, mainly include transmission and distribution equipment, such as transformers, conducting wires, cables, lighting arresters, electrical insulators, and towers.
3. Water supply facilities
   Water supply facilities include public water supply utilities, regional pressurized water supply facilities, water distribution networks, fire hydrants, and valves.
4. Environmental sanitation facilities
   Environmental sanitation facilities include environmental sanitation public facilities, environmental sanitation engineering facilities, and workplaces used by environmental sanitation institutions. Environmental sanitation public facilities refer to public toilets, septic tanks, garbage pipes, garbage containers, garbage containers rooms, waste containers, and spittoons. Environmental sanitation engineering facilities are used for collection, transportation, treatment, and consumption of garbage and feces in environmental sanitation work, including garbage and feces wharf, garbage transfer station, innocuous treatment plant (field), garbage storage yard, garbage composting plant, temporary emergency waste storage yard, water supply tap, and vehicle washing station. The workplace used by environmental sanitation institutions refers to the environmental sanitation operation team's places to complete their management and business duties, including the working room for environmental sanitation management, vehicle parking lot, build and repair factory, and rest places for cleaners.
5. Cleaning and maintenance of ground transportation, water supply, power supply, and other supporting facilities in the mining area
   Recommendations for supporting facilities maintenance:

   (1) The drainage facilities such as drainage pump station and sewage pump station shall be painted once a year and cleaned regularly.
   (2) Newspaper kiosks, report boxes, mailboxes (barrels), and other post and communication facilities shall be renovated every six months and cleaned regularly.
   (3) Power distribution room (box) and power switching station shall be oiled once a year and cleaned regularly.
   (4) The greening installations around the buildings shall be renovated every six months and cleaned regularly.
   (5) Waste transfer stations, public toilets, and other environmental sanitation facilities shall be painted or oiled once a year, with periodic cleaning.

Onsite evaluation or supporting materials:

(1) Check the relevant design data of the mine.

(2) Check the ground transportation, water supply, power supply, environmental protection facilities, environmental sanitation facilities, and other systems on the mining area's general ground layout.
(3) Sort water supply, power supply, ground transportation, and other mine facilities' data into documents and upload them as an appendix.

3 Living facilities

Standard score: 15

Instructions: Staff dormitory, canteen, bathhouse, toilet, and other facilities should be well-equipped, clean, and tidy with standardized management. Five points are deducted for each non-conformance

Assessment method: Check the site

Explanations: This article reflects how living facilities guarantee production.

The mining area's living facilities, including staff dormitory, canteen, bathhouse, and toilet, should be regularly disinfected to keep the internal room clean and tidy. The specific requirements are as follows:

1. Staff canteen

    (1) The floor, walls, tables and chairs, doors and windows, glass, food windows should be wiped clean without dirt and leftovers.
    (2) The cutting board, cage cloth, cooking utensils, knives, and tableware shall be cleaned and disinfected as required.
    (3) Raw and cooked meat, vegetables, and utensils should be separated.
    (4) Flies, rats, and toxic substances are strictly forbidden in the canteen. The food warehouse goods should be placed away from the ground, and the moldy and expired food should be disposed of in time.
    (5) The staff should wear clean and tidy clothes without any makeup and be polite.
    (6) The gas utensils should be placed and used according to the regulation, and the electrical equipment should be installed reasonably.
    (7) The canteen surrounding environment shall be cleaned timely without sundries.

2. Staff bathhouse

    (1) The indoor floor, walls, doors and windows, shower equipment should be clean and sanitary without any dirt, cigarette end, and damage.
    (2) Clothes boxes in the changing room should be clean and sanitary, without damage, graffiti, and phlegm marks.
    (3) Heating facilities should be in good condition, without air, water, and air leakage.

(4) Anti-rodent, anti-mosquito (fly), and fire-fighting equipment are complete.
(5) The laundry and drying facilities should be clean and in good condition. The installation of electrical appliances, power distribution boards, and lines should meet the standard, and the illumination facilities should be well equipped.
(6) The surrounding environment shall be clean and tidy, and the garbage and coal dust shall be cleaned up in time without residue.

3. Staff dormitory

(1) The personal belongings should be placed in order, and the floor should be free of dirt and cigarette ends.
(2) No graffiti and phlegm marks on the outside walls. Moreover, there should be no weeds, leaves, cigarette butts, cigarette boxes, waste paper (boxes), packaging bags, and leftovers outdoors.
(3) The outdoor garbage shall be put into the dustbin and transported away in time.
(4) No clothes and bedding drying outside.
(5) Heating facilities should be in good condition, without air, water, and air leakage.

Onsite evaluation or supporting materials:

(1) The site conditions of the dormitory, canteen, bathhouse, toilet, and other facilities should be inspected from various aspects, angles, and levels (with photos). The appearance of the canteen, operation room, dining room, and other places' sanitation should be inspected; the size and facilities of the bathhouse should be examined; the staff dormitory should be inspected for its clean appearance and indoor hygiene; the examination of the toilet should focus on sanitation conditions and whether there is flushing water.
(2) The canteen's appearance, canteen operation room, canteen dining room, the interior of the bathhouse, staff dormitory's appearance, the interior of the staff dormitory, and the public toilet. Each place mentioned above should be taken one picture for further inspection.
(3) Relative account records.

4 Plates in the production zone

Standard score: 15

Instructions: (1) Operation notice boards, instruction boards, circuit diagram boards, and other plates shall be set in the production zone as required. Three points are deducted for each non-conformance. (2) The size, shape, and color of the plate shall meet the requirements. Three points are deducted for each non-conformance

Assessment method: Check the site
Criteria or standard: *Plates* (GB/T 13,306) and *Mine Safety Signs* (GB 14,161)

Explanations: This article is the requirement for operation notice boards, instruction boards, route map boards, and other plates in the mining production zone, reflecting the production facilities' service capacity.

1. Places where plates are required
   (1) Operation notice boards, instruction boards, and route map boards
      1) Underground locations and workplaces, including shaft horse headgate, yard, roadway intersection, blasting equipment warehouse, oil depot, fan station, and refuge chamber.
      2) Locations and workplaces on the ground, including wellhead, ground road, substation, power distribution room, hoist engine room, surface explosive magazine, compressor room, main ventilator room, tailings dam, waste rock yard, protective slope, flood discharge ditch, and other safety facilities.

   (2) Safety signs
   Safety signs shall be set at road intersections, ground substations, wellhead, power distribution room, hoist engine room, main ventilation room, slope, petrol station, oil depot, and other alerting areas.

   (3) Graphical signs for environmental protection
   Graphical signs for environmental protection should be set at solid waste storage and deposit sites.

2. Requirements for plates and signs setting
   The size, shape, and color of the plate shall meet the requirements. There are three types of plates in mines.

   (1) Plates
   Plates (GB/T 13,306), applied to product nameplates, operation notice boards, instruction boards, circuit diagram boards, design data charts, and safety signs of various mechanical and electrical equipment, instruments, and components, specifies the type and size, marking, technical requirements, inspection methods, inspection rules, packaging, storage and transportation of plates.

   (2) Mine safety signs
   Mine Safety Signs (GB 14,161) and *Safety Signs and Guideline for the Use* (GB 2894–2008) regulate primary safety information transmitting signs in various mines, including symbols, names, and setting places (See Table 1).

Table 1 Main safety information transmitting signs

Number	Sign	Name	Setting place
2–1		No fireworks	Places with no fireworks
2–2		Entering the well(pit) after drinking is strictly forbidden	Wellheads and pits with access
2–3		No open flames operation	Places with no open flames operation
2–4		Starting the machine is prohibited	Mechanical and electrical equipment forbidden to start

(3) Graphical signs for environmental protection
Graphical Signs for Environmental Protection Solid Waste Storage (Disposal) Sites (GB 15,562.2–1995) regulate the types, symbols, colors of graphic signs for environmental protection in general solid waste hazardous waste storage and disposal sites, as well as use and maintenance.

3. Building (structure) railings, signboards, and other ancillary facilities shall be painted once a year and cleaned regularly.

Onsite evaluation and supporting materials: Leachate and its treated discharge

(1) Check whether there are plates and whether the plates meet the standard at no less than five places, such as operation prompt boards, instruction boards, circuit diagram boards, and safety signs.
(2) One notice board, one instruction board, one circuit diagram board, and one safety signboard at different positions should be provided for inspection.

5 Fixation management

Standard score: 15

Instructions: The equipment and materials shall be managed in order and appropriately placed. Five points are deducted if the equipment and materials are thrown and managed in disorder at each place

Assessment method: Check the site

Explanations: This article is the requirement for the equipment and materials placed in the production area, reflecting the production and living-related equipment and materials' management ability.

1. Fixation management
   Fixation management, a science to study the relationship among people, things, and places in enterprises production activities, is the specific management of things and a comprehensive application and supplement of other professional management in the production site. Fixation management includes the fixation of space and time. Space fixation should deal with the relationship among human beings, objects, and places; time fixation should deal with human beings, objects, and time.

   Space fixation: Space fixation is an essential means to realize the production site's standardization and scientific management. Through organizations, the unnecessary part in the production process will be eliminated, the production site conditions will be improved continuously, and the scientific use of sites will be realized to obtain high space-utilizing efficiency benefits.

   Time fixation: Time fixation promotes the combination of human beings and objects through rectification to obtain benefits with high time-utilizing efficiency.

   The space fixation management integrates the core content "arrangement, rectification and literacy" into the enterprise fine management activities. Its content is relatively abundant, involving the design and optimization of the "person-object-place" three-dimensional spatial position relationship in the whole enterprise. Enterprises can also separate this work for more detailed research and implementation.

   Generally, there are three states between objects and users (human beings) in an enterprise.

   State A: Users and objects reach a compelling combination. People can use things directly, conveniently, quickly, and effectively.

   State B: Users and objects cannot combine effectively. People cannot use things directly, conveniently, quickly, and effectively.

   State C: There is no link and combination between users and objects. The objects are in an unusable and unutilized state.

   Fixation management can continuously eliminate state C, improve state B, and maintain state A. Fixation management should first design and optimize the "person-object-place" three-dimensional spatial position relationship and make full use of color, indicator boards, electronic screens, and other information media to express the information such as where the object is, where the object is going, the amount of the object, what kind of processing state it is in, whether it is in good condition, whether it is dangerous, and the area's overall fixation status, to facilitate people's search and identification, reduce work errors, and improve work efficiency.

2. Fixation management in the production zone
   Equipment and materials in the production zone need to adopt fixation management to ensure orderly placement and stacking.

   (1) The production equipment, facilities, materials, domestic garbage, and waste materials brought by the lifting from underground wells must be recycled to the designated area within the shift time.
   (2) The production materials stored in the production zone must be placed in order according to the classification, and the site should be exact during processing and manufacturing.
   (3) Newly purchased production materials such as yellow sand, foam mud, stone, cement, and stone shall be stored in strict conformity with the regulations and shall not occupy the road. The square piles shall be in order.

3. Requirements for material stacking on the construction site

   (1) General requirements
       1) The stacking of building materials depends on the number of materials, using time, supply, and transportation conditions. Materials with large consumption, long using time, and convenient supply and transportation should be imported in stages and batches to reduce the storage yard and warehouse occupation.
       2) The stacking of various tools, components, and materials at the construction site must be placed according to the general layout plan.
       3) Select a convenient location to facilitate transportation, loading, and unloading. Secondary handling should be reduced.
       4) The terrain should be high, solid, and flat, and the backfill soil should be compacted in layers. Drainage measures should be adopted to meet the requirements of safety and fire prevention.
       5) Materials should be stacked according to the varieties and specifications, with prominent plates indicating the name, specifications, and origins.
       6) All kinds of materials must be stacked in order
   (2) Stacking of semi-finished products of primary materials
       1) Large-scale tools should be aligned at one end.
       2) Reinforcement bars should be stacked in order and padded with square timber. Avoid being placed in a humid place or exposed to rain.
       3) The bricks shall be stacked into square piles with 0.5 m away from the edge of the trench pit. Control the piles' height to prevent collapse.
       4) Stones should be stacked in a square according to different size specifications.
       5) All kinds of formwork should be stacked in order according to the specifications, the ground should be flat and solid, and the stacking height should not exceed 1.5 m; large formwork should be placed on

the specially designed storage shelf, and two pieces of large formwork should be face-to-face. When stored on the construction floor, it should satisfy the self-stability angle and have reliable anti-dumping measures.

6) The stacking site of concrete components should be solid and flat. Concrete components should be stacked according to specifications and models with correct chocks supporting. The chocks of multi-layer components should be aligned up and down with a not very high position. The concrete wallboard should be equipped with an inserting frame, which should be welded or bound firmly to prevent collapse.

Onsite evaluation and supporting materials:

(1) Check the equipment and materials' stacking at no less than four places on the site. Four points will be deducted if the equipment and materials are not placed in order in one place.
(2) Take pictures of fixation management at the cave mouth, warehouse, maintenance workshop, warehouse, parking lot, bicycle shed, stockyard, office, and an underground working face.

6 Solid waste stacking

Standard score: Seven

Instructions:

(1) Three points are rewarded for solid waste having fixed stacking sites
(2) Four points rewarded for solid waste stacking sites complying with regulations

Assessment method: Check the site

Criteria and standards: *Standard for Pollution Control on the Storage and Disposal Sites for General Industrial Solid Wastes* (GB 18,599), and *Standard for Pollution Control on Hazardous Waste Storage* (GB 18,597)

Explanations: This article is the requirement for waste stacking management in the mining area, reflecting solid waste management in the mining area.

Waste refers to the articles and materials useless for a particular activity under specific conditions and discarded. Broadly speaking, it includes solid, liquid, and gaseous wastes. Narrowly speaking, it usually refers to solid waste or waste containing a large amount of solid; however, "useless, discarded" is a relative concept. It does not mean that the discarded waste has lost its use-value completely. When the technical and environmental conditions change or the activity mode, method, and purpose change, the discarded waste may become raw materials, fuel, or consumables for production. Waste recycling is based on this theory.

Solid waste refers to the solid, semi-solid, and gaseous substances placed in containers that have lost their original use value or have not lost their use value but have been abandoned or discarded in production, life, and other activities, as well as those substances included in the solid waste management according to laws and administrative regulations.

Special stacking places should be set up for solid waste from production and living, and its operation, construction, and management should comply with regulations.

Special stacking place for solid waste should conform to *Law of the People's Republic of China on the Prevention and Control of Solid Waste Pollution, Regulations on the prevention of geological disasters* and other regulations on safety, environmental protection and monitoring, and comply with the requirements from *Standards for Pollution Control on the Storage and Disposal Sites for General Industrial Solid Wastes* (GB 18,599—2001), and *Standard for Pollution Control on Hazardous Waste Storage* (GB 18,597—2011).

Standard for Pollution Control on the Storage and Disposal sites for General Industrial Solid Wastes (GB 18,599—2001) has clear regulations on the solid waste stacking site, environmental protection, operation management, site closure, and site monitoring. The specific requirements are as follows:

1. The storage and disposal sites are divided into types I and II.
   Type I general industrial solid waste: According to the *Solid Waste-extraction Procedure for the Toxicity of Solid waste-Retroflexion Method* (GB 5086.1—1997) and *Solid Waste-extraction Procedure for Leaching Toxicity-horizontal Vibration Method* (GB 5086.2—1997), the concentration of any pollutant in the leaching solution does not exceed maximal emission concentration regulated by *Integrated Wastewater Discharge Standard* (GB 8978—1996). Moreover, the pH of the leaching solution ranges from six to nine. The industrial solid waste where this leaching solution comes from is considered type I general industrial solid waste.

   Type II general industrial solid waste: According to the *Solid Waste-extraction Procedure for Toxicity of Solid Waste-retroflexion Method* (GB 5086.1—1997) and *Solid Waste-extraction Procedure for Leaching toxicity-horizontal vibration method* (GB 5086.2—1997), the concentration of one or more pollutants in the leaching solution exceeds maximal emission concentration regulated by *Integrated Wastewater Discharge Standard* (GB 8978—1996). Furthermore, the pH of the leaching solution is beyond the range of six to nine. The industrial solid waste where this leaching solution comes from is considered type II general industrial solid waste.

   The storage and disposal site for type I general industrial solid waste is type I storage and disposal site, abbreviated as type I site. The storage and disposal site for type II general industrial solid waste is type II storage and disposal site, abbreviated as type II site.

2. Requirements

(1) Common requirements for type I and type II sites
   1) The selected site should meet the requirements of the local urban and rural construction plan.
   2) The site location and the distance between the site and surrounding population should be determined according to the environmental impact assessment and approval from the administrative department of environmental protection, which can also be used as the basis for planning and control.

      In the environmental impact assessment of the general industrial solid waste storage and disposal site, the air pollutants such as leachate and dust should be considered first. Comprehensively evaluate the impact on the surrounding environment, residents' health, daily life, and production activities according to the types of environmental functional areas. Then decide the reasonable spatial relationship between deposit sites and the sensitive objects, such as permanent residential area, agricultural land, surface water body, highway, main traffic road (national or provincial road), railway, airport, military base.
   3) The storage and disposal site should be built on the foundation meeting the bearing capacity requirements to avoid the influence of foundation settlement, especially the uneven or local settlement.
   4) Avoid the fault, fault fracture zone, karst cave area, and natural landslide or debris flow area.
   5) Beaches and flooded areas below the highest water level of rivers, lakes, and reservoirs are prohibited for site selection.
   6) Nature reserves, scenic spots, and other areas that need particular protection are prohibited for site selection.

(2) Other requirements for type-I sites
   Abandoned mining pits and collapse areas are preferred.

(3) Other requirements for type-II sites
   1) Type-II sites should avoid the main recharge area of groundwater and aquifer of drinking water.
   2) Type-II sites should be built on a foundation with good anti-seepage performance. The distance between the surface of the natural foundation layer and groundwater level shall be no less than 1.5 m.

3. Environmental protection

(1) Common requirements for type-I and type-II sites
   1) The construction type of storage and disposal site must comply with the type of general industrial solid waste.
   2) Set special evaluation of storage and disposal sites in the environmental impact assessment of construction projects. For those rebuilt

and extra-served sites, the environmental impact assessment should be conducted again.

3) Take dust prevention measures in storage and disposal sites.
4) Build diversion channels around the storage and disposal sites to prevent rainwater runoff from entering the storage and disposal site, leading to increased leachate volume and landslides.
5) Design leachate collection and drainage facilities.
6) Construct embankments, dams, and retaining walls to prevent general industrial solid wastes and leachate loss.
7) Prevent foundation sinking, especially uneven or partial settlement.
8) Take measures to prevent coal gangue's spontaneous combustion with a sulfur content more than 1.5%.
9) Set up the graphic signs of environmental protection in the storage and disposal sites to strengthen the supervision and management according to *Graphical Signs for Environmental Protection Solid Waste Storage (Disposal) Sites* (GB 15,562.2—1995).

(2) Other requirements for type-II sites

1) When the natural foundation layer's permeability coefficient is greater than 1.0×10^7 cm/s, natural or artificial materials should be used to construct the anti-seepage layer. The thickness of the anti-seepage layer should be equivalent to the clay layer's anti-seepage performance with the permeability coefficient of 1.0×10^7 cm/s and the thickness of 1.5 m.
2) If necessary, leachate treatment facilities should be designed to treat leachate.
3) At least three groundwater quality monitoring wells should be built around the storage and disposal site to monitor the groundwater pollution caused by leachate. The first well is located upstream of the storage and disposal site along the groundwater flow direction as a control well. The second well is located downstream of the storage and disposal site along the groundwater flow direction as a pollution monitoring well. The third well is located around the storage and disposal sites where the diffusion effect is most likely to occur as a pollution diffusion monitoring well.

4. Operation management

(1) Common requirements for type-I and type-II sites

1) After the storage and disposal site is complete, the environmental protection administrative department that examined and approved the environmental impact report (form) must check and accept the sites before producing and using.
2) It is forbidden to mix hazardous wastes and domestic wastes at the storage and disposal sites of general industrial solid wastes.

3) The leachate of storage and disposal site can be discharged after the water quality meets *the Integrated Wastewater Discharge Standard* (GB 8978–1996); Air pollutants can be discharged after they meet the unorganized emission requirements in the *Integrated Emission Standard of Air Pollutants* (GB 16,297–1996).
4) The storage and disposal site shall establish an inspection and maintenance system, regularly inspect and maintain facilities such as dike, dam, retaining wall, and diversion channel, and timely take necessary measures to ensure regular operation in case of possible damage or abnormality.
5) Users of storage and disposal sites shall establish a file system and record the type, quantity, and the following information of general industrial solid wastes in detail for long-term storage and reference at any time. The information includes inspection and maintenance data of various facilities and equipment, observation and disposal data of foundation subsidence, collapse and landslide, and monitoring data of leachate, water pollutants, and air pollutants.
6) The graphic signs of environmental protection at storage and disposal sites shall be inspected and maintained following the provisions in the *Graphical Signs for Environmental Protection Solid Waste Storage (Disposal) Sites* (GB 15,562.2–1995).

(2) Other requirements for type-I sites
Mixing type-II industrial solid wastes are prohibited.

(3) Other requirements for type-II site
1) Inspect and maintain the anti-seepage works regularly. Monitor the groundwater quality regularly. Take necessary measures if the anti-seepage function declines. The groundwater quality shall be evaluated according to the *Quality Standard for Ground Water* (GB/T 14,848–2017).
2) Inspect and maintain the leachate collection, drainage facilities, and leachate treatment facilities regularly. Monitor the quality of leachate and its treated discharge water regularly. If the collection and drainage facilities malfunction or the treated water quality exceeds the *Integrated Wastewater Discharge Standard* (GB 8978–1996) or the local pollutant discharge standard, necessary measures should be taken in time.

5. Site lockdown

(1) Common requirements for type-I and type-II sites
1) When the storage and disposal site expires or no longer undertake new storage and disposal tasks for some reason, the storage and disposal site shall be closed or sealed off. Before closing or sealing off a site, a plan must be submitted to the local administrative department

of environmental protection at or above the county level for approval, and pollution control measures shall be taken.

2) When the site is closed or sealed off, the surface slope shall not exceed 33 degrees. There needs a step with a width of no less than 1 m, a slope of 2 -3% and strength capable of withstanding rainstorm scouring every three to five meters elevation.

3) After the site lockdown, continue the maintenance and management until it is stable to prevent the subsidence and cracking of the overburden layer, which would result in a leachate volume increase. Furthermore, prevent landslides and other accidents caused by the instability of general industrial solid waste piles.

4) After the site lockdown, set up signs to indicate the closing time and notices when using the land.

(2) Other requirements for type-I sites

The surface should be covered with a layer of natural soil to restore vegetation when closing, and its thickness depends on the particle size of solid waste and the type of plants.

(3) Other requirements for type-II sites

1) Cover two soil layers on the site's surface to prevent direct exposure of solid wastes and infiltration of rainwater into the pile body. The first layer is a barrier layer, covered with 20–45 cm thick clay and compacted to prevent rainwater from penetrating the solid waste heap. The second layer is the covering layer, with natural soil for plant growth. Its thickness depends on the plant species.

2) After site lockdown, the monitoring system of leachate and its treated discharge water should continue to operate normally until the water quality is stable. The groundwater monitoring system should continue to operate normally.

6. Monitoring

(1) Pollution control projects

a. Leachate and its treated discharge water

Select characteristic components of general industrial solid waste as control projects.

b. Underground water

Before the storage and disposal site is put into use, the items regulated in the *Quality Standard for Ground Water* (GB/T 14,848–2017) shall be taken as the control items. The characteristic components of the solid waste stored and disposed can be selected for the control projects during use or after closing and sealing.

c. Atmosphere

The control project of the storage and disposal sites refers to particulate matter. For the storage and disposal site of spontaneous-combustion coal gangue, particulate matter and sulfur dioxide are the control projects.

(2) Monitoring
   a. Leachate and its treated discharge water
      1) Sampling points: Discharge ports.
      2) Sampling frequency: Once a month.
      3) Determination method: Refer to the *Integrated Wastewater Discharge Standard* (GB 8978–1996)
   b. Underground water
      1) Sampling point: Underground water quality monitoring well.
      2) Sampling frequency: Before the storage and disposal site is used, the background level should be monitored at least once. During the operation, or after the site is closed, sampling shall be carried out once a period according to the dry, normal, and wet periods
      3) Determination method: Refer to the *Standard Examination Methods for Drinking Water—General Principles* (GB/T 5750.1—2006).
   c. Atmosphere
      1) Sampling point: Refer to *Appendix C* in the *Integrated Emission Standard of Air Pollutants* (GB 16,297–1996).
      2) Sampling frequency: Once a month.
      3) Determination method: Refer to Table 2.

The *Standard for Pollution Control on Hazardous Waste Storage* (GB 18,597–2001) regulates hazardous waste stacking and disposal. The storage zones of explosives, power oil, flammable materials, and other hazardous wastes should be reasonably designed according to the type and quantity of waste. According to the characteristics of the waste in each zone, anti-seepage, anti-corrosion, fire prevention, lightning protection, dust prevention, and other measures should be adopted. If there is an equipment maintenance workshop in the industrial site, a special temporary storage room for hazardous wastes should be established. The temporary storage room should be sealed off, and the ground should be implemented impermeable treatment. Cofferdams or diversion ditches and collection wells should be set around the storage facilities.

Onsite evaluation and supporting materials:

(1) Solid waste stacking design of mine.

Table 2 Determination method for air pollutants

Project	Determination method	Source
Particulate matter	Gravimetric method	GB/T 15,432—1995
Sulfur dioxide	(1) Formaldehyde absorbing-pararosaniline spectrophotometry	HJ 482—2009
	(2) Tetrachloromercurate(TCM)-pararosaniline method	HJ 483—2009

(2) Contracts with the unit with hazardous waste storage qualification and certificates from hazardous waste storage units.
(3) Photos of fixed stacking places for solid wastes.

7 Solid-waste management

Standard score: Eight

Instructions: Eight points are rewarded for standardized solid-waste stacking operation and pollutions control, including seepage and domestic waste management

Assessment method: Check the site

Criteria and standards: *Standard for Pollution Control on the Storage and Disposal Sites for General Industrial Solid Wastes* (GB 18,599), and *Standard for Pollution Control on Hazardous Waste Storage* (GB 18,597)

Explanations: This article is the requirement for solid waste management.

The governing laws and regulations include *Law of the People's Republic of China on the Prevention and Control of Solid Waste Pollution, Standards for Pollution Control on the Storage and Disposal Sites for General Industrial Solid Wastes* (GB 18,599–2001), *Standard for Pollution Control on Hazardous Waste Storage* (GB 18,597–2001), *Standard for Pollution Control on the Hazardous Waste Landfill* (GB 18,598–2001) and *Standard for Pollution Control of Hazardous Waste Incineration* (GB 18,484–2001).

The mine should establish a particular solid-waste management system, including solid-waste classification, collection and storage, hazardous waste identification marks, and solid-waste treatment, disposal, inspection, and record.

Solid-waste operation and management should comply with the operation and management requirements in the *Standards for Pollution Control on the Storage and Disposal Sites for General Industrial Solid Wastes* (GB 18,599–2001).

Onsite evaluation and supporting materials:

(1) Special solid-waste management system.
(2) Hazardous-waste identification marks.
(3) Check whether there are seepage and leakage and whether there are domestic wastes mixed. If so, take photos.
(4) Solid-waste management ledger.

8 Domestic-waste disposal and utilization

Standard score: 20

Instructions: (1) Five points are rewarded for domestic waste collection in a fixed place in the mining area (including underground mine). (2) Five points for domestic waste classification and reasonably determining the scope, variety, requirements, collection, and transportation methods of garbage. (3) Ten points for domestic waste treated harmlessly by itself or a third party, with supporting materials provided

Assessment method: Check the site

Explanations: This article is the domestic-waste disposal and utilization, reflecting domestic-waste management in the mining area.

The regulations and standards for reference include *Technical Guide for Domestic Waste Treatment* (City construction [2010] No.61), *Classification and Evaluation Standard of Municipal Solid Waste* (CJJ / T 102–2004), *Technical Specification for Domestic Waste Collection Station* (CJJ 179–2012), Construction Standard for Domestic Waste Collection Station (JB 154–2011), *Technical Code for Municipal Solid Waste Sanitary Landfill* (GB 50,869–2013), *Standard for Pollution Control on the Landfill Site of Municipal Solid Waste* (GB 16,889–2008), *Technical Code for Linear System of Municipal Solid Waste Landfill* (CJJ 113–2007), *Technical Code for Projects of Municipal Waste Incineration* (CJJ 90–2009).

Article 43 in the *Law of the People's Republic of China on the Prevention and Control of Solid Waste Pollution* stipulates that local governments at or above the county level shall accelerate establishing a domestic waste management system of classified release, collection, transportation, and treatment, to realize domestic waste classification system's effective coverage.

1. Waste collection

   The domestic waste collection point is a component of the domestic waste collection system. It is located at the front end of the waste collection system, with the characteristics of large quantity, wide distribution, and difficult management. Domestic waste collection point directly affects the mining area's living environment and the mining area's construction and management level.

   According to the definition of *Terminology Standard for Urban Appearance and Environmental Hygiene* (CJJ / T 65–2004), the domestic waste collection point is the place set up for waste collection. Its function is to collect garbage, store it briefly for transportation.

   Domestic waste collection points should be convenient for putting in and transporting garbage. The domestic waste collection point's appearance is also significant because it is usually located at the side of the road or in the building and yard.

   (1) The collection points shall be approved by the local administrative department of environmental health.
   (2) Domestic-waste collection points should be convenient to put garbage in without affecting the mining area's sanitation and landscape environment.
   (3) Domestic-waste collection points should cater to the container's movement and rotation and facilitate mechanized collection and transportation.
   (4) Maintain reasonable protection distance and set greening or blocking measures to reduce visual and olfactory pollution.

(5) The ground where the collection point stands should be flat and hardened.

2. Waste classification
The significance of centralized waste classification is as follows.

(1) Waste classification conducive to garbage recycling.
Waste classification can improve resource utilization. Compared with exploiting natural resources, recycling metals, paper, plastic, and glass from waste has the advantages of a high recycling rate, simple treatment, low cost, and less pollution. Moreover, food, plants, and fabrics in the garbage can be composted to produce organic fertilizer; bricks, tiles, and lime soil can be processed into building materials. Many resources can be reused in the garbage. Solid wastes are mixed to be garbage and separated into resources. Sorting and collecting wastes and harmless treatment will transfer them into considerable recycling resources.
(2) Waste classification conducive to waste reduction.
Sorting and collecting waste can effectively reduce the total waste amount. Urban waste is divided into renewable waste (waste paper, waste plastic, waste metal, waste glass products, and waste furniture), non-renewable waste (meal waste, and peel) and hazardous waste (waste batteries and waste lamps). According to statistical analysis, in recent years, the proportion of renewable waste in urban garbage has increased rapidly, which is beneficial to domestic waste classification and recycling. Classification and recycling will transfer waste into useful resources, thus reducing the total waste amount. Waste reduction can reduce the landfill load, treatment cost, land occupation, and the adverse impact on the atmosphere, soil, surface water, and groundwater. The full removal of plastics, rubber, and electronic products, is also of great benefit to waste incineration. It reduces harmful waste-gas generation, such as dioxin, which will alleviate the serious secondary pollution problem of waste incineration in China.
(3) Waste classification is conducive to waste harmlessness.
Waste sorting collection can improve the harmless treatment rate. At present, the main waste treatment methods in China are landfill, incineration, and composting. Although the landfill method has the advantages of low treatment cost and simple method, it may cause groundwater resources secondary pollution. The incineration method occupies less land and has a good reduction effect, but it is easy to produce chemical substances with high toxicity. The composting method can make organic matter in the waste decompose into fertilizer to improve the soil, but the composting scale should not be too large, and the long-term use of compost causes groundwater quality deterioration and soil hardening. A lot of domestic wastes can be recycled. Suppose we start from the initial link of waste generation, carry out classification and recycling, select waste carefully, and find the useless wastes for treatment. In that case, we can reduce

the total urban garbage amount, save money, reduce the terminal treatment pressure, alleviate the garbage accumulation situation, and reduce the environmental pollution in the treatment process, which significantly improves the harmless treatment rate of urban waste.

3. Domestic waste self-harmless treatment
Domestic waste self-harmless treatment refers to using advanced technology to reduce the impact of garbage and its derivatives on the environment to reduce waste discharge and achieve resource recycling. There are three kinds of domestic waste self-harmless treatments: incineration, sanitary landfill, and composting. The domestic waste treatment process should be based on the actual local situation, waste composition, urban economic development, people's living habits, as well as meteorological, hydrological, and geological conditions. Fulfill the harmless waste level in a short time with the least investment.

(1) Incineration
If the harmful gases, organic matter and slag produced during incineration are directly discharged into the environment, they will also lead to pollution. Hence, it is necessary to send the waste into the incinerator for centralized treatment (some domestic wastes contain less organic matter, so coal and garbage can be mixed and incinerated in a certain proportion).
Waste incineration in an incinerator is a thermochemical treatment method using high temperatures to destroy and change waste's composition and structure to reduce the volume and make it harmless. After incineration, the volume can be reduced by 80- 90%, and the harmful substances and bacteria are basically eliminated.
Based on incineration in an incinerator, the heat can be used to generate electricity. Waste incineration for power generation solves the environmental pollution caused by open-air waste incineration and effectively utilizes the energy contained in the garbage, serving a double purpose.
(2) Sanitary landfill
With the continuous development of waste treatment technology, the waste treatment is transiting from stacking to landfill, and landfill technology is gradually sanitary, economical, and efficient. The ash, brick and tile, ceramics, and other materials in the garbage and various residues produced after the garbage treatment must be disposed by landfill.
(3) Composting
Composting is an organic fertilizer that is made of animal and plant remains and excrement containing fertilizer, soil, and mineral matter, fermented and decomposed under high temperature and humidity. Compost is a kind of fertilizer with a long history. To make compost, we must collect appropriate materials, such as straw, stems and vines, weeds, tree leaves, or livestock manure. Mix them properly, and add the proper amount of calcium cyanamide to promote fermentation. Finally,

cover with broken mats, rags, straws, or plastic clothes to avoid fertilizer loss.

Onsite evaluation and supporting materials:

(1) Photos of waste-collection-points. The enterprise will get no score if the wastes are dumped everywhere in the production area, management area, and living area without waste collection facilities.
(2) Photos about waste collection.
(3) The domestic-waste harmless treatment scheme or the third party treatment contract.

9 Main road condition
Standard score: 15
Instructions: Eight points are rewarded for the main roads in the mining area meeting the regulations with smooth, dense, and rough surfaces; seven points for good maintenance
Assessment method: Check the site
Criteria and standards: *Code for Design of Roads in Factories and Mines* (GBJ22)

Explanations: This article is the requirement for the mining area's road surface, reflecting the mining area's transportation guarantee.

1. Roads classification in factories and mines

(1) The highway outside the factory refers to the external highway connecting mining enterprises with national roads, urban roads, stations, and ports, or the connecting roads between the scattered workshops (branches) and residential areas.
(2) The factory roads are the factory's internal ones (or ports, commercial warehouses, open-pit mine machine repair sites, and mine wellhead sites; the same below).
(3) Open-pit mines transportation roads are the roads for mining trucks in open-pit mines. They often serve as the auxiliary roads (including the ground auxiliary roads of mines), leading to blasting material storehouse, water source, general substation, tailings dam. The transportation roads can be divided into arterial roads, branch roads, connecting line roads, and auxiliary line roads. Arterial roads refer to the common roads from each open-pit mining bench to ore unloading point or waste dump (waste rock yard). Branch roads refer to the roads connecting the mining bench or waste dump (waste rock yard) with the arterial roads, or the roads from a mining step directly to the ore unloading point or waste dump (waste rock yard). Connecting line roads refer to other roads connected with arterial

roads and branch roads in open pit mines. Auxiliary line roads lead to the auxiliary plants (workshops) and various auxiliary facilities in the mining area.

2. Main road specifications in the mining area
According to the requirements in the *Code for Design of Roads in Factories and Mines* (GBJ 22–1987), special arterial roads in the mining area should have sufficient strength and good stability, and road surface should be flat, dense, with proper roughness.

 (1) Asphalt and cement concrete can be used for roads in the production area with a large amount of dust.
 (2) Cement-concrete precast-block pavement or block stone pavement can be used for road sections with underground buried pipelines and frequent excavation and maintenance.
 (3) Block stone pavement can be used for road sections with large slopes or smaller circular curves.
 (4) Block stone pavement or low-grade pavement can be adopted for the road where tracked vehicles are often used.
 (5) Roads in salt lake mines and alpine area mines can take corresponding measures according to specific conditions [not required by the *Code for Design of Roads in Factories and Mines* (GBJ 22–1987)].

3. Pavement induration's function
Pavement induration is to cover hardened layers such as bitumen and concrete on the formed pavement. Its functions are as follows.

 (1) Improve transportation. Vehicle travel is faster and more convenient after pavement induration.
 (2) Change the mining environment. Hardened roads tend to cleaner and more beautiful, which is different from the dirty, disordered, and poor appearance in the past.
 (3) Reduce transportation energy consumption.

4. Road maintenance
Roads' common problems include peeling, sanding, freezing and thawing, potholes, cracks, edge failures, and broken corners.
"Highway maintenance" refers to road maintenance and repairs. Maintenance focuses on the whole maintenance process from the road completion, and repairs focus on the damaged parts. Road maintenance is divided into daily maintenance, regular maintenance, special maintenance, and improvement engineering.

 (1) Daily maintenance
 Daily maintenance is frequent daily operations for each road component (including ancillary facilities) according to the annual requirements to maintain the original good state and service level.

1) Pavement and other parts cleaning.
2) Minor damage repair and sporadic replacement of facilities.
3) Mowing and branches pruning.
4) Snow and ice removal in winter.
5) Emergency treatment for occasionally interrupted traffic.

(2) Regular maintenance
Regular maintenance is a relatively large-scale maintenance operation that can be programmed and carried out within the service life.
1) Auxiliary facilities improvement.
2) Pavement wearing course renewal or repair.
3) Repair of pavement markings, culverts, and ancillary facilities.
4) Repaint metal bridges.

(3) Special maintenance
Special maintenance restores the seriously deteriorated road condition to the original state.
1) Strengthen and rebuild the damaged pavement structure.
2) Repair damaged subgrade and culvert.
3) Take measures to prevent and control external factors damage to the highway, such as stabilizing slope, preventing collapse, adding retaining walls, improving drainage facilities, preventing water damage, preventing avalanche, and cutting down trees.

(4) Improvement works
1) Improve congested road sections and increase traffic capacity.
2) Correct crown and superelevation to improve sight distance.
3) Adjust intersections and entrances to eliminate accident-prone points for safety.
4) Take anti-noise measures.
5) Expand and rebuild buildings and other facilities.
6) Add roadside rest areas to improve the highway service level.

Onsite evaluation and supporting materials:

(1) The access road, the connecting road in the office area, and the road from the office area to the production area. Each road surface should be taken one picture for further inspection.
(2) Check the road maintenance ledger.

10 Road cleanliness
Standard score: Ten

Instructions: The internal roads or special roads in the mining area have no scattered objects, or effective measures should be taken to clean up the scattered objects in time. Five points are deducted for each non-conformance

Assessment method: Check the site

Explanations: This article is the requirement for the internal roads or special roads cleanliness in the mining area, showing the mining area's social image.

1. Cleaning range
   The cleanable area (including road green space) between the walls on both sides of the road, the graffiti on buildings and attachments on the ground, and the cleaning of construction waste and debris along the road.
2. Road cleaning requirements
   (1) The road cleaning operation shall meet the corresponding road level standards in the road section. The fast lane, slow lane, and sidewalk shall be cleaned thoroughly without missed sweeping, dead corners, peel, paper scraps, plastic bags, leaves, cigarette butts, gravel, dust, and other wastes.
   (2) Keep the road surface, stones, and wall corners free of water, and do not block the maintenance holes.
3. Road cleaning measures
   (1) Timely collect and clean the cleaning garbage and transport it on time.
   (2) Clean the garbage containers in time to keep the surrounding area free of debris.
   (3) Timely remove the road pollution and cooperate with the sprinkler to actively wash the road surface.
4. Operation site management
   To strengthen each operation site's effective management, conduct the onsite inspection and track the implementation of cleaning objectives, the management department should timely correct unqualified services to ensure the road cleaning operation is carried out smoothly. The specific contents are as follows:
   (1) Work out the cleaning work schedule, and list the cleaning items and time arrangement according to the cleaning work's overall scope and procedure details at the operation point. Arrange personnel posts and implement a post responsibility system following the cleaning work schedule.
   (2) Cleaning posts need to be arranged according to the post scope, work items, health standards, and the requirements of "five fixed" posts (fixed personnel, fixed posts, fixed quantity, fixed time, and fixed standard).
   (3) All cleaning operations must be completed according to the cleaning procedures and sanitary standards.
   (4) Road cleaning must comply with the operating procedures. Implement safety operation management measures to ensure a safe and civilized operation.
   (5) Ensure the cleaning quality and efficiency and improve the management system. Management personnel and employees must fill in the "personnel operation registration book" and "daily inspection registration book" according to the requirements of "road cleaning operation standards and assessment rules."

(6) In the operation process, we must take good care of the public facilities and equipment within the road scope.
(7) During holidays, more road cleaning workers should be arranged according to each road section's conditions.

Onsite evaluation and supporting materials:

(1) Photos of falling objects on the pavement of access road, connecting roads in the office area, and roads from the office area to the production area.
(2) Check the road cleaning measures.

11 Mine cleanliness
Standard score: 20
Instructions: The mining area shall be clean and sanitary with no garbage or waste rocks in the production area and management area. There shall be no gas running, water inrush, dripping water, and leakage in pipelines on the production site. Five points are deducted for each unqualified item
Assessment method: Check the site

Explanations: This article is the requirement for the production and administration area's sanitation, showing the mining area's social image.

Running, inrush, dripping, and leakage refer to the phenomenon of gas running and liquid leakage caused by improper management and operation in the process of storage and transportation.

1. Production- and administration-area sanitation
Ensure that the road and sidewalk are clean and free of floating dust, sundries, oil pollution, garbage, waste rock, ponding, snow, stain, and sediment.
2. No gas running and liquid leakage in the living area

(1) Gas running and liquid leakage control in the water supply and drainage system
Inspect the water supply and drainage system pipes weekly. Timely repair and replace the pipelines and control valves if there is gas running and water leakage.
When the equipment's water tank and pipe fittings in the water supply and drainage system have gas running and liquid leakage, they shall be repaired in time.
The underground galvanized pipe needs to be replaced every three years.
(2) Gas running and liquid leakage control in the oil supply system
Inspect the relevant devices weekly. In the case of gas running and water leakage, timely maintenance shall be carried out.
According to the equipment requirements, waste oil should be replaced regularly and treated in a centralized way.

When the equipment's pipeline system, control valve, and solenoid valve have gas running and liquid leakage, they shall be repaired in time.

Relevant departments shall inspect the gas running and liquid leakage from time to time and record them in the form of environmental and safety inspection records.

Onsite evaluation and supporting materials:

(1) Check at least six places' equipment and facilities' actual situation on sites, such as pipelines, control valves, solenoid valves, pools, and transportation equipment.
(2) Photos of actual onsite situation.
(3) If dust accumulation is severe on the road surface and both sides, and the sewage and wastewater flow everywhere, it cannot be scored.

12 Construction and maintenance of buildings and structures in the mining area

Standard score: 20

Instructions: (1) Twelve points are rewarded for no temporary buildings or abandoned buildings in the production area, management area, and living area; four points are deducted for each non-conformance

(2) Eight points for timely maintenance, repair, or painting of buildings and structures in the mining area. Two points will be deducted for each obvious damage, aging without any repair and maintenance measures

Assessment method: Check the site

Explanations: This article is the requirement for the buildings and structures management in the mining area, showing its social image.

1. Relevant laws and regulations

   There are no clear legal restrictions on illegal buildings. Generally speaking, it refers to changing the original building and other useful methods outside the prescribed scope to achieve their interests. *General Principles of the Civil Law of the People's Republic of China* and *Urban and Rural Planning Law of the People's Republic of China* have relevant requirements.

   (1) *General Principles of the Civil Law of the People's Republic of China*

   Article 83 In the spirit of helping production, making things convenient for people's lives, enhancing unity and mutual assistance, and being fair and reasonable, neighboring real estate users shall maintain proper neighborly relations over such matters as water supply and drainage passageway, ventilation, and lighting. Anyone who causes obstruction or damage to his/her neighbors shall stop the infringement, eliminate the obstruction, and compensate for the damage.

(2) *Urban and Rural Planning Law of the People's Republic of China*
Article 64 Article 64 If a construction project is proceeded without obtaining the planning permission on a construction project or by violating the provisions of the planning permission on a construction project, the competent department of urban and rural planning of the people's government at or above the county level shall order it to stop construction. If it is still possible for the construction entity or individual to take measures to eliminate the impact on the implementation of urban and rural planning, the department shall order it or him to correct within a certain time limit and impose a fine of not less than 5% the construction cost but not more than 10% the cost; if it is impossible to take measures to eliminate the impact, the department shall order the construction entity or individual to dismantle the building or structure within a certain time limit and confiscate the real objects or the illegal gain. It may also impose a fine not more than 10% of construction cost.

Article 65 Where a construction project is proceeded within a township or village planning area without obtaining the planning permit for rural construction or by violating the provisions of the planning permit for rural construction, the people's government of the township or town shall order the construction entity or individual to stop construction and correct within a certain time limit. If the entity or individual fails to do so within the time limit, the buildings or structures should be dismantled.

Article 66 Where a construction entity or individual commits any of the following behaviors, the competent department of urban and rural planning of the people's government of the city or county where it is located shall order it to dismantle the building within a certain time limit, and may impose a fine not more than the cost of the temporary construction project:

1) Making temporary construction without approval.
2) Making temporary construction beyond the approved content.
3) Failing to dismantle the temporary building or structure after the approved time limit expires.

2. Temporary construction
Temporary buildings refer to the structures and other facilities that must be demolished within the time limit (generally two years), as well as the structures and other facilities with steel, reinforced concrete, cement, brick, wood, stone and other durable materials due to engineering needs which must be demolished within two years (such as some temporary exhibition rooms of Shanghai World Expo).

The construction of temporary buildings shall also be approved by the planning, land, and construction departments, and the service life shall not exceed two years.

The temporary building adopts a disassembly structure, which requires using more times, assembling and disassembling quickly, less loss, and low economic index. The geometry of the structure should be as simple as possible. For every temporary building, the unified modulus of plane and section is selected to minimize the types of components. The building blocks are assembled into different spans, different areas, and different uses by the building block principle.

3. The facade and public facilities of buildings (structures) in the mining area shall be clean and tidy

   (1) The facade and public facilities cleanliness should meet the following standards:
   (2) No obvious stain and dust.
   (3) No damage and surface falling off.
   (4) No obvious old, paint mottling and serious fading.
   (5) No graffiti, depiction, and post.
   (6) No other unclean appearance, which would affect the city appearance and environment.

4. The buildings and structures style should be in harmony with the surrounding environment

   The external facades of building attachments and structures that may affect the building must be coordinated with the main building's color, shape, and design style. Pay attention to the following aspects.

   (1) The decorations on the buildings and structures facade should maintain the original color and style. According to the building facade actual situation, simple, bright, and harmonious decorative lines or color blocks can be appropriately used to increase the building's three-dimensional and hierarchical effect.
   (2) For older buildings, the facade decoration should combine reinforcement and waterproof, and the facade of the building can be decorated appropriately. The top of low buildings (three stories and below) should also be cleaned and painted.
   (3) Buildings decorations should pay attention to color harmony and collocation. The buildings' facades are mainly light colors with other colors properly configured. It is not suitable to use large-scale gorgeous colors and dark colors to paint buildings. At the same time, the color of a building or a street should consider unity and harmony. Colors should not be too much or too chaotic.
   (4) The short-term and long-term, single building, and overall effect should be fully considered.
   (5) For large-area glass curtain walls and buildings decorated with special materials, the building facade shall be cleaned according to the regulations, and the stains and dirt on the building surface shall be removed to ensure the facade clean and tidy.
   (6) The functional zoning and the buildings' appearance and structures should be neat, smooth, beautiful, and intact.

5. Buildings and structures management

   Illegal construction mainly refers to the illegal land occupation and illegal building. Illegal construction seriously affects the community residents' lives and infringes on the rights of other citizens. It not only affects the beauty but also leaves many security risks.

   Specifically, unauthorized construction means that some individuals use public fire escape passage to construct illegal buildings beyond the normal standard without the planning department's approval or the fire department's evaluation. Private connection refers to connecting wires and water pipes without application.

6. Building (structure) facade in the mining area shall be clean and tidy

   Building (structure) facade in the mining area shall be kept clean and tidy. The suggestions are as follows.

   (1) Facades made of glass or ceramic tile should be cleaned once a year.
   (2) Spray-painting type facade should be renewed every three years.
   (3) In principle, when buildings (structures) facade in mining area is painted or redecorated, the color, shape, and design style shall be maintained the same as the original buildings (structures). Moreover, coordination with the surrounding environment shall be noted.

Onsite evaluation and supporting materials:

(1) Appearance photos of main buildings in the office area and auxiliary production area.
(2) Photos of illegal construction.

2 Mine Afforestation

13 Afforestation in the mining area

Standard score: 20

Instructions: The afforestation area in the mining area should be fully covered with no large surface soil area exposed. Five points are deducted for each non-conformance

Assessment method: Check the site

Explanations: This article is the requirement for the greening coverage, showing the mining area's social image.

Afforestation is not equal to ecological restoration, but one of the ecological restorations. Its function is to improve the environment and maintain the ecological balance in the mining area. Precisely, afforestation supplements oxygen in the air, absorbs harmful gases in the atmosphere, prevents dust and wind, reduces noise, sterilizes, improves microclimate, purifies water quality, and keeps the ground dry. Afforestation is to plant and maintain plants by artificial methods.

1. Functions of afforestation

   Afforestation can improve environmental sanitation and play a variety of roles in maintaining ecological balance.

   (1) Replenish the air with oxygen. Green plants absorb carbon dioxide in the air and release oxygen in photosynthesis. Sixty percent of the oxygen in the air is produced by forest and green land.
   (2) Absorb harmful gases in the atmosphere. Many plants have a strong ability to absorb and filter harmful gases in the atmosphere. Greening can effectively reduce the nitrogen oxides in automobile exhaust, thus reducing the atmosphere's ozone and preventing photochemical smog formation. Besides, some leaves can also absorb lead, cadmium, and arsenic from the atmosphere.
   (3) Dustproof. Some leaves and stems have villi, and others can secrete mucus or oil. Therefore, they can intercept, filter, adsorb, or adhere to various particles suspended in the atmosphere. According to statistics, the amount of atmospheric particulate matter in the area with good afforestation is 50–70% less than that in the non-greening area. After being washed by rain and snow water, the plants can also recover their dust collection ability. Grassland also has a dust prevention function. The grass itself can fix the soil on the land surface and prevent secondary dust. Thanks to the rough surface, grassland can make the particles near the ground stagnate on the grassland.
   (4) Windbreak. Afforestation is one of the effective measures to prevent wind, especially in the dense forest. Since tree trunks, branches, and leaves can block the airflow, airflow speed will slow down after passing through the forest. The weakening degree is related to trees' height, quantity, and species and forest width. For example, when the air flows through the forest with a width of 120–240 m, the wind speed can almost be reduced to zero, so afforestation has become a critical wind protection measure.
   (5) Noise reduction. The results show that the noise intensity can be reduced by 10–13 dB for a medium forest, 18–20 dB for a dense forest, and 4–7 dB for a common garden in the street. Therefore, trees and plants can play the role of sound insulation wall and muffler. Trees and shrubs are planted on both sides of the road and around houses to reduce noise pollution.
   (6) Sterilization. Some plants can secrete mucus to retain bacteria in the air and release fungicides with the bactericidal effect. Therefore, planting more trees and flowers can reduce some airborne diseases.
   (7) Improve the microclimate in a region. Plants improve microclimate by blocking solar radiation and adjusting the temperature. The green area's

solar radiation temperature is about 14% lower than that in the non-greening area. Afforestation can reduce the degree of heat and severe sun exposure, making the area warm in winter and cool in summer.

(8) Purify water quality. Trees have the function of absorbing dissolved substances in water. The results show that after the colored, odorous, turbid, and bacteria-containing sewage flows through the forest, the color decreases, the odor weakens or disappears, the transparency increases, and the bacterial content decreases significantly, which significantly reduces the environmental pollution caused by the sewage.

(9) Keep the ground dry. In marsh areas and areas with high groundwater levels, the ground is often very wet, bad for health. Trees have a strong ability to absorb and store underground water. Therefore, afforestation has become an effective measure to keep the ground dry.

Afforestation is not equal to ecological restoration, but one of the ecological restorations. Its function is to improve the environment and maintain the ecological balance in the mining area.

2. Afforestation layout and area

Code for General Layout Design of Industrial Enterprises (GB 50,187–2012) has the following provisions.

(1) The industrial enterprises' green space rate should be controlled within 20%, and that of reconstructed and expanded industrial enterprises should be controlled within 15%. Industrial enterprises with special requirements such as product safety can be excluded, and local planning and control requirements can also be implemented according to construction projects' specific conditions. The afforestation layout shall meet the following requirements: (a) the non-building section and scattered open space in the mining area shall be fully utilized for greening; (b) the site under the facilities such as pipe rack, trestle, overhead line, and above the underground pipeline belt shall be used for greening; (c) the requirements for production, maintenance, transportation, safety, health, fire prevention, lighting, and ventilation shall be met. The mutual influence between buildings, structures, and underground shall be avoided; (d) water cooling facilities' cooling effect should not be hindered.

(2) The following sections are key areas for afforestation: (a) both sides of the main road into the factory and the main entrance and exit; (b) the administrative office area of the enterprise; (c) the production workshop, device and building area with high cleanliness requirements; (d) the production workshop, device and storage yard that emit harmful gas, dust, and high noise; (e) the production workshop and buildings exposed to the western sun; (f) the section washed by rain; (g) places around the living service facilities in the mining area; (h) inside the wall of the main road near the town.

3. Afforestation coverage rate in the mining area

   (1) Afforestation area refers to the bare land that does not affect production and can be afforested. It does not include hardened ground, building coverage area, process unit area, space between oil storage tank group and fire lane, inside fire dike, protective wall, and other afforestation-forbidden areas. Greening areas include plant land areas, garden pavilion, pool, pond, and other land areas.
   (2) The mining area's greening coverage rate refers to the percentage of the area afforested to the green area.
   This article requires that the greening mineralization coverage rate reach one hundred percent. All the greening areas should be afforested, without dead corners and bare surface (mud) soil. Bare bedrock and hardened pavement are not within the greening scope.
   Large area with surface soil exposure needs an assessment unit and joint enterprise evaluation.

Onsite evaluation and supporting materials:

(1) Check the key afforestation layout section at no less than six different places.
(2) Provide site photos once a large area of topsoil is discovered.

14 Special main roads greening and beautification requirements
Standard score: Ten
Instructions: Ten points are rewarded points for access roads in the mining area, the internal roads in the office area, and the roads from the office area to the production area set as following indications
(1) Isolation greening belt should be set up according to local conditions.
(2) If there are no greening conditions objectively, make billboards or slogans
Assessment method: Check the site

Explanations: This article is the requirement for particular main roads greening and beautification in the mining area, showing its social image.

The greening isolation belt is used to isolate the vehicle and beautify the environment. The greening belt between the two sides of the motorway and bike lane is called highway side zoning in municipal roads.

1. Relevant requirements

   (1) Isolation greening belt shall be set on both sides of the mine road with afforestation space, and the setting shall comply with Article 10.3 in *Technical Specifications of Eco-environmental Protection and Reclamation for Mining* (HJ 651–2013).

(2) According to the *Code for Design of Roads for Factories and Mines* (GBJ22-87), afforestation should be implemented on both sides, middle belts, intersections of the road, parking lots and other ancillary facilities. Different kinds of trees should be planted alternately during greening. For arid and water-deficient areas, the mining area's special roads can be beautified according to local conditions.

2. Isolation greening belt functions

(1) Reduce noise interference and prevent environmental pollution. The trees with strong absorption capacity should be selected scientifically to reduce the emission pollution and purify the air.
(2) Separate the opposite traffic flow and significantly reduce vehicle collision possibility.
(3) The greening belt provides space for setting up signs, traffic indicator lines, street lights, and other traffic auxiliary facilities, reducing the driving difficulty, and improving driving safety.
(4) Greening belts can better improve the road landscape, and green facilities with a certain height can prevent glaring.
(5) The greening separation belts on both sides of the special road can afforest the mining area and beautify the road appearance and improve the environment.

3. Roads without greening conditions
For roads without greening conditions, installing shelter walls (boards), billboards or slogans, and other beautification on both sides of the road can reduce the visual landscape's adverse impact or reduce driving fatigue.

Onsite evaluation and supporting materials:

(1) Less than three photos of isolation greening belt on access roads, onsite roads, and other main roads
(2) Make billboards, slogans, and other beautifications on both sides of the roads.

15 Afforestation guarantee mechanism

Standard score: Four

Instructions: Four points are rewarded for a long-term afforestation guarantee mechanism, afforestation maintenance plan, and a responsible person

Assessment method: Check the site and search information

Explanations: This article is the requirement for a long-term afforestation guarantee mechanism, showing the mining area's social image.

Arrange full-time personnel or entrust a professional team to carry out greening management for the mining area.

1. Afforestation management in the mining area
   (1) Formulate greening management and maintenance specifications, and guide daily maintenance work according to professional technical requirements.
   (2) Maintain and improve the greening water supply facilities to ensure the irrigation system's normal operation.
   (3) Strengthen the pruning of lawn, flower, and shrubs, and properly reshape the landscape trees to create a good plant landscape.
   (4) Increase reasonable guide instructions and safety warning signs, and improve service facilities.
   (5) Strengthen the comprehensive control of lawn weeds.
   (6) Timely replant dead and missing hedgerows, flowers, and trees to keep the greening space clean and perfect.
   (7) Strengthen the pest prevention and fertilizer and water management control, and ensure that plants are flourishing, healthy, and full of vitality.
   (8) Take protective measures against the severe cold and deal with emergencies in time.
   (9) Improve the overall quality of horticultural workers.
   (10) Establish and improve plant maintenance and management's technical files to facilitate the summary, work analysis, and improvement.
   (11) Strengthen road tree maintenance and pruning.

2. Requirements and standards for afforestation maintenance service in an enterprise
   The daily greening maintenance management includes the whole process of drenching water, opening the nest to cultivate the soil, pruning, fertilization, weeding, edge clearing, pest control, reinforcement, and seedling supplement. Table 3 shows the maintenance frequency and quality standards.
3. Afforestation maintenance assessment standard of a unit
   Table 4 shows the afforestation maintenance assessment standard of a unit.

Onsite evaluation or supporting materials:

Special departments (or individuals) check the management system and maintenance account. Outsourcing investigation on greening maintenance contract.

16 Greening guarantee effect

Standard score: Six

Instructions: Six points are rewarded for reasonable greening-plant collocation, no serious dead branches, and yellow leaves. Two points are deducted for each non- conformance

Assessment method: Check the site

Table 3 Afforestation maintenance frequency and quality standard in an enterprise

Categories	Maintenance content	Frequency				Maintenance standards
		Per week	Per month	Per season	Per year	
Lawn	Lawn mowing		1			The lawn should be smooth and beautiful, with a grass height of 3-5 cm
	Grass trimming		1			The lawn edge junction should be clear and smooth, without a well-defined or long boundary
	Mowing		1			The garbage transfer station designated by the party a should be cleaned up in time, the working site should be cleaned after construction
	Lawn weeding		1			The purity rate of the lawn should reach 95% without any apparent broadleaf weed
	Lawn fertilizing			1		The lawn should be watered once a season, making the green period last over two hundred and forty days in a year
	Lawn watering	3				The lawn should be watered three times per week to keep it hydrated without any wilting situation. The staff needs to pay attention to the weather forecast to save water

(continued)

Table 3 (continued)

Categories	Maintenance content	Frequency				Maintenance standards
		Per week	Per month	Per season	Per year	
	Lawn pest control	1				The lawn needs to be checked once a week. Diseases and pests should be treated in time once they are found; the pest control rate should reach 95%
	Bare loess on the lawn	1				The bare loess on the lawn should be checked once a week and be treated in time. The lawn coverage rate should reach 98%
	Lawn green belt garbage removal	7				The garbage needs to be cleaned up twice a day, making the cleaning rate 95%. Garbage, leaves, and debris are not allowed on the lawn
Trees	Tree trimming			1		There should be no excessive growth of branches, dry branches, rotten logs, over dense branches, and rotten branches on the tree
	Tree scarification			2		The soil around the root should be loose and flat, without moss and hardening soil
	Tree fertilizing				2	The fertilizer should not be exposed to the ground, and the soil should not be in shortage of fertilizer
	Tree weeding		1			There should be no apparent weeds and 30 cm long sprouting tillers

(continued)

Table 3 (continued)

Categories	Maintenance content	Frequency				Maintenance standards
		Per week	Per month	Per season	Per year	
	Pest control		1			The disease and pests control rate should reach 95% without obvious pests. (except termites and forest quarantine pests)
	Tree watering	2				It is necessary to maintain the normal growth of trees without water shortage
Shrubs	Shrub trimming		1			The plants should have a precise shape and be trimmed straight with no serious dry branches and yellow leaves; the new branches' length should not exceed 10 cm; the flowering plants should bloom at a proper time
	Shrub scarification			2		Scarification needs to be carried out twice a season to make the soil loose. The moss and hardening soil should be eliminated
	Shrub fertilizing			1		Shrubs grow well without any shortage of fertilizer and the excessive growth of branches
	Shrub pest control		1			Pests should be treated once they are found. There should be no obvious branches of diseases and pests. Pest control rate should reach 95% (except for termites and forest quarantine diseases and insect pests)

(continued)

Table 3 (continued)

Categories	Maintenance content	Frequency				Maintenance standards
		Per week	Per month	Per season	Per year	
Hedgerow and green wall	Trimming		1			The surface should be straight and trimmed into a precise shape. The side should be vertical with no obvious missing cutting, collapse, and dry branches. The feet of the hedgerow and green wall should be neat
	Fertilizing			1		The hedgerow and green wall should be treated with the proper amount of nutrients. There should be no shortage of fertilizer and no excessive growth of branches
	Scarification			2		Hedgerows edge is neat with loose soils
	Watering	2				It is necessary to maintain the normal growth of hedgerows without water shortage
	Pest control		1			Pests should be treated once they are found, without obvious branches of diseases and pests. The pest control rate should reach 95%
Ground cover plants of the flower bed	Trim		1			The flower bed edge should be neat and beautiful, treated with reasonable pruning without obvious residual flowers. The flower bed should have clear layers, which could contribute to the sense of three-dimension

(continued)

Table 3 (continued)

Categories	Maintenance content	Frequency				Maintenance standards
		Per week	Per month	Per season	Per year	
	Fertilizing			1		The plants in the flower bed should be treated with the proper amount of nutrients. There should be no shortage of fertilizer and no excessive growth of branches
	Watering	2				It is necessary to maintain normal growth without water shortage
	Scarification			2		Hedgerows edge is neat with loose soils
	Pest prevention		1			Pests should be treated once they are found, without obvious branches of diseases and pests. The pest control rate should reach 95%

Table 4 Afforestation maintenance assessment standard of a unit

Categories	Inspection items	Standards
60 points for maintenance standards	Trees	The trees thrive, and their branches and leaves are strong without withering
		The shape of plant growth characteristics should be maintained. Obvious excessive branches, over dense bore branches, dry branches, rotten logs, over dense branches, and rotten branches shall be eliminated
		The pruning effect should be in harmony with the surrounding environment, and the lower edge line should be higher than 2 m
		The tree chamber should be well ventilated. The main and lateral branches should be evenly distributed (except for the situation that the growth of the tree crown is hindered)
60 points for maintenance standards	Trees	There should be no obvious disease and pest branches. The damage rate of diseases and pests should be no more than 8%, and the damage rate of a single plant should be no more than 8%
		There should be no dry branches and yellow leaves that are contrary to the growth characteristics
		There should be no fruit on non-fruiting trees. Their fruits need to be pruned in time
		The trees with flowering branches in the same year should be cut moderately in winter, and some main lateral branches should be reserved
		Typhoon prevention measures shall be taken, and the service center shall be fully cooperated to help complete typhoon prevention work. The slanted trees affected by external force shall be righted within 1–2 days after external force
		Bags at trees' head should be reserved at the places where people and vehicles pass. The soil in the bags should be loose

(continued)

Table 4 (continued)

Categories	Inspection items	Standards
		There should be no mechanical damage, random pulling, and hanging phenomenon on the branches. The trees should have no curly leaves, yellow leaves, and abnormal fallen leaves
		For trees whose growths are seriously affected, pruning and other remedial measures should be taken. However, the green tissue on the trunk should be exposed outside with suggestive signs
		New trees, especially ancient trees, should be protected by shading
		The soil at the root should be loose and flat, without moss, hardened soil, exposed fertilizer, and lack of fertilizer
		It is necessary to maintain the normal growth of trees without water shortage
		Tree holes or damages found in the trunk should be repaired in time
		Timely desilting and drainage should be carried out for the trees with accumulated water inside the trunk
	Modeling plants and shrubs	Within the range of $1000m^2$, dry shrubs do not exceed $2\ m^2$ accumulatively, and a single dead area does not exceed $0.5\ m^2$
		The trees thrive, and their branches and leaves are strong without dead trees
		There should be no obvious disease and pest branches. The damage rate of diseases and pests should be no more than 8%, and the damage rate of a single plant should be no more than 8%
		The tree should not have serious dry branches and yellow leaves. And there should be no seedling lack, dead seedlings, and bare loess
		The plants should have a precise shape and straight angle and be trimmed straight; the new branches should be cut once their lengths exceed 10 cm
		Flowering shrubs should be reserved to ensure blooming, and branches should not be too messy

(continued)

Table 4 (continued)

Categories	Inspection items	Standards
		Overwintering cutting should not hinder the view, and some main lateral branches should be retained
		It is necessary to maintain the normal growth of shrubs without water shortage. There should be a proper amount of fertilizers without fertilizer shortage and excessive growth
		The shrubs' edge should be neat with loose soils
	Hedgerow ground cover and flower cluster	There should be no obvious disease and pest branches. Pests and diseases damage rate does not exceed 5% per 100 square meters
		When many plants are planted in patches, the contour lines should be neat and layered, with inner plants higher than outer plants
		The hedgerow and flower cluster side should be trimmed neatly and beautifully, and the lines of the single flower bed should be clear; the side of the hedgerow is trimmed into an inverted trapezoid
60 points for maintenance standards	Hedgerow ground cover and flower cluster	There should be no dry branches, dry leaves, large construction waste, and bare loess in the hedgerow and flower cluster. If the branch length is over 10 cm, it should be pruned
		There should be no weeds in the hedgerow and flower bed, and the grassland should not extend into the hedge and flower bed; the grass edge should be trimmed
		The edge of the uncut flower bed and hedge should be neat with loose soils
		The edge of the grass should be cut neatly. Weeds should not enter the flower cluster
		The growth should be exuberant. Dead seedlings and missing seedlings should be replaced in time

(continued)

Table 4 (continued)

Categories	Inspection items	Standards
		Over dense flowers, clusters should be divided in time, and aged flower clusters should be replanted in time
		It is necessary to maintain normal growth without water shortage. There should be proper amounts of fertilizers without fertilizer shortage and excessive growth
		There are suggestive marks for the flowers that are renewed and replanted
	Lawn	Grass thrives, with dark green leaf color. The lawn's overall level and height are below 4 cm
		The lawn coverage rate should reach 98%
		The examined weed rate should be lower than 3%, with a purity of 97%
		The lawn should be repaired within three days after damaged by an external force
		The total exposed area of loess in every 1,000 m^2 green space does not exceed 2 m^2, and each area is not more than 0.5 m^2
		There should be no obvious disease and pest branches. Pests and diseases damage rate does not exceed 3% per 100 square meters
		There should be no obvious broad-leaved weeds on the lawn, and the weed rate of the lawn should not exceed 3%
		The grass should be flat, without potholes, rat holes, sharp protrusions. All wires and pipes in the grassland should not be exposed, and the aged or protruding lawn should be thinned
		The lawn should be arranged a proper amount of water without wilting
		The grass edge should be trimmed neatly, with a clear and smooth edge junction and clear boundary. Grass should not grow to the curb and flower cluster

(continued)

Table 4 (continued)

Categories	Inspection items	Standards
	Overall	There is no lack of water and fertilizer and no bare loess
		There should be no garbage, stones, dry branches, and leaves in the hedgerow and flower cluster
		The site shall be cleared after the completion of the greening operation. The greening garbage shall be collected in time and shall not be stored overnight
		There should be no obvious dust on the important parts of the blade
		Greening killing must not use the national banned highly toxic pesticides
		During the mechanical operation, the operators shall wear protective equipment and have onsite operation identification; the operation time shall not affect the owner's normal life
		The site shall be cleared after the greening operation. The greening garbage shall be timely cleaned up to the designated garbage transfer station and shall not be stored overnight
		The greening tools and water pipes for site operation should be placed in order and should not affect the owner's normal life and passage
Sixty points for maintenance standards	Overall	There should be no obvious dust on the leaves of important parts
		Save water and close the faucet in time. Water pipes should have no obvious damage, leakage with no long flowing water
		The plants in the aerial layer and under the dense crown should meet the needs of growth
		The plants in the shade, such as the overhead layer and tree crown, should meet the needs of growth
	Vertical greening	Climbing plants grow luxuriantly, with edges trimmed neatly. They are not supposed to affect the traffic and the normal operation of facilities and equipment such as infrared monitoring

(continued)

Table 4 (continued)

Categories	Inspection items	Standards
		The coverage rate in the growing period should reach 95%
Twenty-five points for operation requirements	Indoor greening	Plants thrive without pests, dry branches, and dead branches. The leaf surface should be bright and clean without dust
		The flowerpot and bottom plate should be clean
	Trees	The maintenance frequency should be strictly implemented
		Tree pruning should be guaranteed twice a year (before typhoon and after winter)
		For palm plants and plants with weak growth, organic fertilizer should be applied once a year, and compound fertilizer should be applied once a year. The depth of fertilization should be no less than 30 cm
		Trees that affect the owner's life must be pruned in time
	Shrubs and flower bed	The maintenance frequency shall be strictly implemented
		The water supply of flower bed should be guaranteed 2–3 times per week
		Pruning should be carried out 2–3 times per month in peak season and 1–2 times per month in a slow growth period
		Organic fertilizer should be applied once a year for isolated shrubs, and the depth of fertilization should not be less than 20 cm
		The flower bed should be fertilized once a season; the plants for viewing flowers should be reserved to ensure that the flowers are flourishing and the branches are not too messy
	Lawn	The maintenance frequency shall be strictly implemented
		It is pruned once or twice a month from May to October every year, and once a month from November to April in the following year

(continued)

Table 4 (continued)

Categories	Inspection items	Standards
		Ensure lawn fertilization 1–2 times/month
		Lawn watering should be guaranteed 4–5 times per week from May to October every year, and 3–4 times per week from November to April in the following year (if it is rainy weather, the watering can be reduced)
	Replenish seedlings and grass	Seedlings and grass replenishment should be finished within three days
	Weeding	3–4 times per year
	Pest control	According to different varieties and seasons, pest control should be carried out once a month with a complete killing
Twenty-five points for operation requirements	Pest control	Green seedlings should have no common diseases with the damage rate of diseases and pests under 8%, and the damage rate of the individual plant under exceed 8%.
		Diseases and pests branches should be pruned in time and treated correctly.
		The personnel should master the knowledge of killing pests. The drug selection and proportion should meet the requirements
Fifteen points for behavior standard	Code of conduct for maintenance personnel	The staff should wear work clothes and a badge during working hours. The badge should be worn in a conspicuous place on the left chest (about 1cm above the middle of men's coat pocket)
		The staff should be friendly, generous, elegant, active, and warm, and polite.
		The staff should take the initiative to make way for others and guide the way in a polite gesture.
		Workers should take the initiative to greet, nod, or say hello when seeing the owner at any workplace.

(continued)

Table 4 (continued)

Categories	Inspection items	Standards
		The staff's hair should have a natural color. The front hair should not suspend below the eyebrow, side hair does not cover ears, and back hair does not touch the back collar. Workers should shave every day.
		The staff should not chat with others and do other things irrelevant to work.
		The staff should be energetic, enthusiastic in work and good condition, using polite words: Hello, please, sorry, thank you, and goodbye.
		Staff must not eat smelly food before going to work. They should keep the mouth clean. Smoking or drinking in the workplace is strictly prohibited to prevent the smell of smoke or alcohol.
		Staff should put relevant signs on the operation site to remind the owners.
		Staff should wear black or dark opaque socks.

Explanations: This article is the requirement for the greening effect, showing the mining area's social image.

The corresponding greening guarantee effect can be achieved by seriously implementing the greening guarantee mechanism. The main effects are as follows: the greening tree species and plants grow well and vigorously. The branches and leaves are robust. There is no dead, no obvious long branches, dead branches, rotten head, over dense branches, rotten branches, and over dense bore branches. The grass edge is cut neatly, and weeds do not enter the flower cluster.

Take measures to handle the problems in time.

Onsite evaluation or supporting materials:

Check no less than six places at the site. Take photos of the problems discovered.

17 Mine beautification

Standard score: Ten

Instructions: Ten points are rewarded for building parks, gardens, and green lands with the advantage of the natural conditions, topography, and landform

Assessment method: Check the site

Explanations: This article is the requirement for beautification, reflecting the enterprise's image.

Apart from the afforestation, purification, and brightening in the mining area mentioned above, it is necessary to carry out some beautification. The beautification should make full use of the mining area's natural conditions and topography and avoid scribbling and random stacking.

Landscape refers to the complex composed of land, space, and objects on the land.

1. Greenspace

   Green space is the abbreviation of urban green space. Generally, a piece of open space with plants can be called green space. According to the *Standard for Basic Terms of Urban Planning* (GB/T 50,280–1998), urban green space improves the ecology, protects the environment, provides recreational sites for residents, and beautifies the landscape, including public green space, residential green space, and road green space. Green space can be artificial or natural. It is characterized by large area, irregular, and a variety of plants or wild plants.

   (1) Public green space

   Public green space is a kind of recreational green space that meets the sunshine requirements. It is suitable for recreational facilities arrangement and is shared by residents, including residential parks, small parks, group green space, and other block belt green space.

   (2) Residential green space

   Residential green space is a part of residential building land and an important part of residential areas.

In the residential area balance sheet, only the area and percentage of public green space are shown, and residential green space is not included in the index of public green space. Residential green space is 2–3 times larger than that of public green space, and the per capita green space can reach 4–6 m^2. Residential green space is the continuation and supplement of the house's interior space, closely related to residents' daily lives.

(3) Road green space

Road green space refers to the ground used for planting plants and landscaping within the road land scope. It is located at subgrade slope, borrow pit, central partition belt, separation belt, protective belt, square, and sidewalk. Road green space is used to purify the air, beautify the environment, adjust the climate, prevent noise, prevent snow, prevent fire and sunshade, prevent glare, and guide drivers' sight. Road green space has several forms: street trees, avenues, hedges, flower clusters, and strip grass.

According to the relevant provisions in the *Control Index of Land for Industrial Project Construction* (GTZF[2008] No. 24), no green space shall be arranged in industrial enterprises, but a certain proportion of green space can be arranged due to the special needs of production technology, and the green space rate shall not exceed twenty percent.

Green area refers to the land area that can be used for greening with an area of no less than 400 m^2, excluding roof greening, vertical greening, and land covered with soil less than 2 m.

2. Gardens

A flower garden is a green space characterized by ornamental plants, the most common garden type. It can be set up independently or attached to the houses, buildings, and parks. The flower garden is a garden with ornamental trees, flowers, and grass as the main body. It can beautify the environment, provide people with flowers and scenery. The garden area is usually small, but there are many kinds of flowers. Flower beds, flower rims, and flower clusters are used to show rich color and posture. Evergreen plants, lawn and ground cover plants are set as embellishment. For example, the garden with certain ornamental plants as the main body is called a special garden, such as peony garden, rose garden, azalea garden, and orchid garden. The size of the garden is not limited, but the decoration must be delicate.

3. Parks

Park refers to the public green space with a certain range of land use, good greening, and certain supporting facilities in the city for people's recreation.

Onsite evaluation or supporting materials:

Photos of parks, gardens, pools, and green landscapes.

Chapter 5
Resource Development Methods

Resource development must implement "unified planning, centralized mining, large-scale mining, and patch treatment" to integrate mining, renovation, and use. The mining area shall plan the mining layout and total amount in a unified manner, utilize the mining solid wastes on a large-scale, manage the environment timely, and restore the ecosystem's function.

The evaluation results of resource development methods should reflect the level of resource development to environmental protection.

1 Resource Development

18 Mining technology

Standard score: 50

Rating illustration:

★ Suitable for open-pit mining:

① Drilling: Adopting wet, dry (with dust collection) and other rock drilling operations

② Blasting: Adopting millisecond blasting, pre-splitting blasting, and smooth blasting

③ Shovel mucking: Adopting large-scale automatic hydraulic shovel mucking equipment, hydraulic excavators or loaders, self-discharging car, large-scale self-mobile crusher, and other advanced equipment

④ Dumping: In the production period, adopting the technology of step-wise internal dumping to maximize internal dump use and reduce the external land occupation

J. Deng et al., *Interpretation of Green Mine Evaluation Index*,
https://doi.org/10.1007/978-981-16-5433-6_5

Those who meet all requirements will get 50 points. Those not involved will be deemed to meet the requirements. Twenty points are deduced for one item that does not meet the requirements until 50 points are deducted

(For both underground and open-pit mining, one of the main mining methods at the present stage shall be selected for scoring, and the scores shall not be accumulated.)

★ Suitable for underground mining:

① Adopting filling method, water-preserved mining and other technologies
② Reducing large-scale surface subsidence caused by mining
③ Utilizing the goaf to dispose tailings, waste rock, and coal gangue on a large-scale

Those who meet all requirements will get 50 points. Those not involved will be deemed to meet the requirements. Twenty points are deduced for one item that does not meet the requirements until 50 points are deducted

(For both underground and open-pit mining, one of the main mining methods at the present stage shall be selected for scoring, and the scores shall not be accumulated.)

★ Suitable for minerals such as petroleum, natural gas, and geothermal mineral water:

① Adopting electric drilling rig and top drive device;
② Adopting drilling technologies such as optimized, fast drilling and managed pressure drilling;
③ Adopting environmentally friendly drilling fluid and recycling technology;
④ Timely and harmlessly disposing drilling wastes such as drilling mud

Fifteen points are deduced for one item that does not meet the requirements until 50 points are deduced

Evaluation methods: Check data and the site.

[Interpretation] This article is a technical requirement for different mining of underground mines, open-pit mines, oil and gas mines, reflecting the ecological environment protection of mining areas by mining technology.

1. Open-Pit Mining
   Open-pit mining removes the ore coverings to obtain the required minerals or extract useful minerals from the open ground stope. Its operations include perforation, blasting, mining and loading, transportation, and dumping.

(1) Drilling

Wet rock drilling means that the pressure water is sent to the bottom-hole when the rock drill is drilling, inhibiting the generation of rock dust, and wetting, washing, and discharging the rock dust. It is a dust prevention measure adopted in rock drilling.

Dry rock drilling (with dust collection) has an orifice and bottom-hole dust-removing types.

The orifice dust-removing type dry dust-catcher is composed of a dust-removing plug and hose, orifice mask, ejector, and dust filter bag. The mechanism is as follows: The ejector connected with the air supply hose of the rock drill generates a negative pressure of above 400 mmhg, making the dust-removing hose and plug and orifice mask in negative pressure. The dust absorption generated in the hole is sucked into the cloth woven-bag dust filter to complete dust collection and filtering. However, its simple facilities are not suitable for downward holes with a depth above 1.5 m and a large inclination.

Bottom-hole dust-removing type. The bottom-hole dust-removing type dry dust-catcher mechanism is as follows: Firstly, the air ejector's negative pressure in the system is applied to the bottom-hole. The dust sucked out from the bottom-hole enters the barrel of the dry dust-catcher installed with ejector and bag dust filter through the dust guide hose. Finally, dust collection and filtration are completed. Although it is easy to use with strong applicability and applicability, its facilities are complicated.

(2) Blasting

Short-delay blasting (millisecond blasting) is a delayed blasting with a delay interval of several to tens of milliseconds. The blasting time interval of adjacent blastholes is extremely short, which causes the energy-domain generated by each blasthole to affect each other, improving the blasting effect and reducing the hazards of blasting seismic effect, shock wave, and flying rock.

Pre-splitting blasting refers to the transverse crack with a certain width that is blasted along the design contour line before blasting in the main blasting area during rock excavation. It can buffer and reflect the vibration wave of excavation blasting, and control its damage to the retained rock mass, obtaining a flat excavation contour. Pre-split blasting is used on vertical and inclined excavation walls, regular curved and twisted surfaces, and horizontal foundation surfaces. Besides, it is suitable for soft stratum with low stability and control of the excavation contour.

Smooth blasting refers to a controlled blasting technique, in which the contour line after segmented or partitioned millisecond blasting meets the design requirements and the free surface is flat by correctly selecting blasting parameters and construction methods. Besides, the new wall it exploded is smooth and undamaged. Its characteristics are: A row of smooth blasting holes with hole spacing matching the minimum resistance line is drilled on the designed excavation contour line, decouple charge, or other charging structures. After the excavation's main body is blasted, the charge in the smooth blasting holes is simultaneously detonated. Finally, a smooth and flat excavation surface is formed through the smooth blast holes.

(3) Equipment
Hydraulic equipment: Through power components, executive components, control components, auxiliary components, and hydraulic oil, the pressure and mechanical energies of the liquid medium in the system are mutually converted, transmitting motion, force, and torque. The advantages are as follows: The range of transmitted energy is large, realizing step-less speed regulation in a wide range. Smooth operation, especially at low speeds. It is easy to operate with remote control, with automation. Good lubrication of moving parts with long service life.

Commonly used shovel-loading equipment include large-scale automatic hydraulic shovel loading equipment (the electric shovel drives the electric motor to drive the hydraulic pump to work), hydraulic excavators (the diesel engine drives the hydraulic pump to work) or loaders, self-discharging hydraulic car, and large-scale self-mobile hydraulic crusher.

(4) Dumping
During the production period, stepwise mining and internal dumping with stepwise internal dumping technology are adopted. In the early stage, the dump is located outside the mining area for not having dumping space. In the later stage, the space that has been mined should be utilized to maximize the utilization of the internal dump and reduce external land occupation.

(5) Advantages and disadvantages of open-pit mining
Advantages of open-pit mining are as follows: Less restriction of mining space; high labor productivity; Low mining cost; low ore loss and dilution; open-pit mining is safer and more reliable than underground mining; short construction time; good working conditions with safe operations.

Disadvantages of open-pit mining are: Dust and harmful gases endanger human health and ecological environment; open-pit mining requires a larger waste rock yard; climate affects open-pit mining.

The "not involved items" in this article refer to construction techniques without drilling or blasting, such as sand excavation, soil excavation, and coal excavation.

2. Underground Mining
Underground mining refers to extract ore from ore blocks in underground deposits, including orebody development, ore block development, undercut, and stoping. The construction of green mines adopts mining methods such as the filling method to reduce surface subsidence and rock displacement, protecting

surface water and groundwater, and ensuring that resource development is given priority to ecology.

(1) Filling method
The filling method fills the mined-out area with filling material to support the surrounding rocks on two sides, prevent the rock from moving, and control the ground pressure. With the mining face's advancement, filling the mined-out area with filling material is the filling mining method.

The filling method should implement the policy of "simultaneous extraction and excavation, and excavation in advance" and formulate an organized plan to improve construction and safety management. Meanwhile, intelligent and mechanized mining supervision is encouraged to ensure the project quality of shaft and drift excavation and the operators' personal safety. The underground mining method in actual mining should be selected according to the mineral type, surrounding rocks' geological conditions, technical deposit conditions, and economic factors to meet the requirements of safety, economy, high efficiency, and quality.

Underground coal mining should strengthen clean mining technology, promote underground gangue's filling technology, control pollutants generation from the source, and reduce environmental pollution and damage. For extra-thick coal seams with thickness greater than 5 m, the mining method of comprehensive mechanized top coal caving shall be adopted. Considering the relationship between coal and gas, coal mines that can extract gas should use coal and gas's coordinated mining technology to utilize coalbed methane comprehensively.

Underground mining methods of metal mines should be selected according to the shape, inclination, thickness, useful components′, distribution of the ore, and the physical, mechanical and chemical properties of the surrounding rock and ore. Besides, the management of opening-out, development, under-cut, and stoping should refer to the mining design plan, adopting the filling mining method instead of the caving mining method. Limestone mines should promote underground mining to protect the ecological environment. However, if there is a solid direct roof-to-floor and the roof's overall thickness is sufficient, the open-field mining method is adopted.

A. Advantages of filling mining

1) Function of emission reduction.
The significance of filling mining technology is beyond the mining scope. There are mines without waste disposal in the world, such as the Grund lead–zinc mine in Germany. It prepares the paste filling materials with the whole tailings after flotation and the crushed stones after gravity concentration to backfill the downward filling drift, without tailings discharged to the tailings pond. The primary condition for this mining is to backfill industrial solid wastes into the ground. The treatment

of industrial solid wastes is the primary problem in mine design, as well as the best way to solve the environmental pollution in mining areas.

2) Function of eliminating ground displacement and subsidence. Filling technology can quickly and effectively fill the mined-out areas, supporting the top rock strata and preventing the deformation and displacement of surrounding rocks. The new working face formed by filling creates conditions for subsequent operations, shortening the mining cycle, and filling with the total productivity improvement. Besides, filling mining technology can prevent large-scale terrane movement and realize mining under water bodies and buildings, protecting the surface from damage and maintaining the original ecological environment.

3) Function of lean and damaged mining and prospective resource protection.
Filling mining technology can be used for complex and changeable ore bodies such as horizontal ore bodies, gently inclined ore bodies, steeply inclined ore bodies, and branched composite ore bodies. The technology achieves the following aspects: The pillar recovery rate and ore grade of large and thick bodies are improved to maximize the ore recovery. Mines that need to exploit the lower part or chassis rich ore preferentially realizes "mining rich and protecting poverty," avoiding the destruction and waste of mineral resources. The waste rock and ballast, and tailings contain useful substances that cannot be preserved by current technology. They are challenging to be chronically stored on the ground, and filling is reliable.

B. Problems to be considered in filling mining
The protection of filling areas is related to the protection level and objectives. The selection of filling areas and filling mining scheme should combine the *Mining Geological Environment Protection and Land Reclamation Plan*, and consider the following aspects:

1) In the "three under" areas (under buildings, railways, and water bodies), selecting the mining method according to the protection level of *Buildings, Water Bodies, Railway and Main Roadway Coal Pillar Setting and Coal Mining Specification*.
2) In environmentally sensitive areas, the damage degree determines the protection levels and mining methods.
3) The filling and mining methods in other areas shall comprehensively consider relevant factors.
4) The filling boundaries are calculated by the influence angle of rock-soil movement and the distance between the filling layers and protected areas' surface. Meanwhile, the protective belts' width (the safety zones with accidents in protected areas or

construction boundaries) is determined by the protection levels of *Buildings, Railways, Water Bodies, and Main Roadway Coal Pillar Setting and Coal Mining Specification.*

5) Giving priority to gangue (waste rocks, tailings), construction wastes, and other solid wastes to fill the mined-out area, avoiding secondary pollution.
6) The land reclamation plan should consider the environmental impact after filling and mining.
7) The non-permissive subsidence of underground mines shall be mined by filling, making the gangue does not exit the pit or the ground piled without gangue.

C. Security assurance

Mining should strengthen the observation of surface movement and rock, railway, and building deformation. Data research on surface subsidence, subsidence speed, deformation value, curvature value, influence scope, as well as three zones of caving, fissure, and bending can grasp the rule for ensuring safety.

Green mining based on cleaner production technology can achieve small disturbance, nonhazardous, and less pollution.

D. Metal mine

Filling mining can reduce ore dilution, loss rate, rock movement, and surface subsidence and improve ore grade. The filling method should be preferred for underground mining of non-ferrous metal mines, which can protect the pillars, control the mining intensity, and strengthen deformation monitoring. Besides, the protection of surface stability reduces the damage of underground mining to the surrounding environment.

Features of filling mining:

1) Avoiding farmland damage and building relocation on the ground. After the mined-out area is filled, the ground will not collapse.
2) Reducing the construction investment and reclamation costs of tailings ponds. Backfilling the mined-out area with tailings and discharging less or no tailings can reduce or not built tailings pond with land area reduction.
3) Protecting the mine environment. The surface will not collapse, reducing the land occupation or pollution of the tailing pond.
4) Safe and full use of resources.

   Filling mining methods include overhand cut and fill stoping, underhand cut and fill stoping, drift-and-fill stoping, waste-lifting stoping, sacked fill stoping, sublevel open stoping, or block open stoping.

E. Coal

The filling mining method is to fill waste gangue and other noncombustible into the goaf to control roof falling, which can safely manage the roof and dispose waste in large-scale. Filling mining methods of coal mines include:

1) Solid filling mining. Fill gangue, fly-ash, drift-sand, and other solid wastes with coal mining hydraulic support, perforated bottom dismountable conveyor, filling loader, and tamping mechanism into the mined area without cementitious materials.
2) Paste filling mining. Stir gangue, fly-ash, cement, and auxiliary materials with water, and then send them into the mined-out area through industrial filling pumps and pipelines.
3) High-water filling mining. Fill the material with high-water or ultra-high water and send it into the mined-out area through the ordinary filling pump and pipeline after being stirred with water.
4) Roadway consolidated filling mining. Stir gangue, fly-ash, cement, and other filling materials with water, and then send them to the filling roadway through industrial filling pumps and pipelines.
5) Separated strata grout filling is a supplementary filling to control surface deformation. Stir the fly-ash, cement, and other materials with water, and drill grouting holes from the ground to the working face's separation zone. Then, the fly-ash slurry is injected to control surface deformation, suitable for controlling surface deformation.

(2) Water conservation mining

Mining engineering changes the original runoff of groundwater, affecting the groundwater flow system. Water conservation mining combines mining methods and engineering methods such as blocking, intercepting, transferring, and filling for ground grouting. The multi-pronged approach realizes the protection of mine water resources and the comprehensive utilization of mining methods.

(3) Advantages and disadvantages of underground mining

Contrary to the advantages and disadvantages of open-pit mining, it will not be discussed here.

In this article, the "not involved items" refer to the mining methods that are not leading geological subsidence, such as strip-partial mining, room, and pillar mining. Coal mines can also be comprehensively assessed for adopting uniform settling or reducing settling measures.

3. Oil and Gas

Oil and gas production extracts oil and gas from underground oil layers. According to whether the wellbore fluids are artificially supplemented with energy, the mining method of oil and gas from underground to the surface is divided into blow-in and artificial lift.

(1) Adopting electric drilling rig and top drive system.

1) Electric drilling rigs (electrically driven drilling rigs) are drilling equipment that uses electricity to drive the drilling tools' lowering and rotation, and drilling fluid circulation, and other drilling equipment. The advantages of electric drilling rigs over traditional mechanical drilling rigs are as follows:

High transmission efficiency. The total transmission efficiency of the common mechanical drilling rig is about 73%, while that of DC electric drilling rig is 86%, and that of AC variable frequency electric drilling rig is 90%.

Good operability (soft characteristics). DC and variable frequency motors realize stepless speed regulation, with little impact on starting and adjusting speed, protecting mud pumps, winches, and rotary tables.

Low fuel consumption. If the drilling depth is 6000 m, the drilling cycle is 120 days, and three wells are drilled every year, the fuel consumption of a DC electric drilling rig is 15–20% less than that of a mechanical drilling rig.

Good borehole quality. The start of the mechanical drilling rig will cause the bottom hole assembly to generate centrifugal force due to the impact acceleration, making the boreholes irregular. On the contrary, the electric drilling rig's stepless speed regulation rig can ensure the boreholes′ quality, which is beneficial to the sidewalls' stability and the cementing quality.

Convenient maintenance management. The diesel of the electric drilling rig runs under optimal working conditions with overload protection, and the overhaul period can be extended. Electric drilling rigs have few wearing parts, and cables can replace chains, belts, and other transmission pairs, protect the overload, reduce vibration and impact, and maintain the working machines' maintenance. The motor and electrical control equipment only need to be dry and clean. The control room and motor have cooling equipment, ensuring the machines' temperature does not exceed 50 °C. Besides, the sealed bearing of the motor is lubricated every three years. However, the maintenance workload of mechanical drilling rigs is large. There are more wearing parts, such as bearings, friction plates, chains, belts, which need to be adjusted and replaced frequently. Results show that the maintenance cost of electric drilling rigs accounts for 30% of mechanical drilling rigs.

A high degree of environmental protection. The electric drilling rigs are equipped with a noise control device that makes the noise lower than the standard 85 dB, and the drillers on the drill floor hear the instructions and alarms. Electric drilling rigs can directly draw power from the industrial power grid without diesel generators when drilling wells in urban.

2) Top drive device: As one of the three major scientific and technological achievements of modern drilling equipment, the top drive device is new drilling equipment that replaces the traditional rotary table-kelly drilling mode, and a high-tech product integrating mechanic, electric, hydraulic, and communication. It moves the drilling power from the rotary table to the faucet, while the drill tools are driven by the top to rotate the drill without the rotary table and kelly bar. Besides, it can use vertical drilling. The drill string can be connected at any point in the derrick to against sticking and resistance during tripping operation, establishing a circulation and rotary drilling tools. Front and back reaming can prevent sticking accidents through sticking and resistance points, which is better than conventional drilling. When the device is applied to difficult and complex drilling operations such as exploration wells, directional wells, horizontal wells, extended reach wells, the economic benefits are particularly significant.

(2) Drilling technologies such as optimized, fast drilling and managed pressure drilling.

A. Optimized fast drilling

The full name of optimized, fast drilling technology is the integrated matching techniques of optimized, fast drilling.

The optimized, fast drilling technology is an integrated drilling technology that uses a long-life screw and high-efficiency PDC bit to optimize the well site, bit, screw, profile, drilling assembly, drilling, and mud parameters. An effective way to increase the ROP and shorten the drilling cycle, include integrated matching techniques such as drilling technology, drill bits, tools, mud, drilling equipment, and drilling instruments.

"Excellent" means high quality and optimal (optimization refers to optimize the well site, drill bit, screw, drilling assembly, and parameters). Meanwhile, "quick" means rapid drilling, well completion, and production organization.

B. Controlled pressure drilling (CPD)

CPD is an adaptive drilling process that controls the annular pressure section in the entire well by determining the downhole pressure window.

Characteristics are as follows:

Keeping the well limit pressure greater than the formation pore pressure prevents the formation of fluid from intruding.

The drilling fluid density is lower than that of conventional drilling, avoiding exceeding the formation fracture pressure gradient.

Use liquid drilling fluid.

Use a closed, pressure-bearing drilling fluid circulation system.

Functions are as follows:

The low pore pressure of formation fracture reduces well kick-lost circulation, which improves the safety of well control to drill deeper open holes.

Make the bushing go deeper to reduce one or more layers.

Promote larger boreholes to reach the target layer.

Reduce lost circulation caused by the annular pressure.

Reduce drilling fluid costs and well control problems caused by seriously lost circulation when large fractures are drilled.

Avoid underground blowouts.

Provide health, safety, and HSE effects, especially at sea with higher requirements.

Reduce non-production operation time.

(3) Environmentally-friendly drilling fluid and recycling technology.

A. Environmentally-friendly drilling fluid.

Environmentally-friendly drilling fluids can protect the environment and meet the drilling engineering needs, such as white oil-based drilling fluids, alkyl glucoside drilling fluids, silicate drilling fluids, synthetic-based drilling fluids, and polyalcohol drilling fluids. However, to meet engineering needs, the drilling fluid system is added with chemical Products. Common environmentally-friendly drilling fluids include environmentally-friendly scale and corrosion inhibitor polyaspartic acid, environmentally-friendly lubricant polyether HLX, environmentally-friendly sulfureted hydrogen scavenger, high-temperature drilling-fluid additive zirconium citrate, and composite ion viscosity reducer PX.

B. Recycling technology

Drilling fluids discharged from drilling is the main pollution source of oil drilling, whose wastes disposal troubles enterprises. The traditional drilling construction mode adopts "the first drilling is started with mixed slurry, the second drilling adds treating agent to transfer slurry inappropriate well section, and the drilling fluid is discharged into the circulation pool to settle and solidify naturally after the drilling is completed." This mode is easy to handle at a low cost. However, with the increase in drilling costs, stringent requirements for environmental protection standards, and enterprise social responsibility manifestation, the cost pressure and environmental dilemmas caused by this model have emerged.

The drilling fluid returned from the well participates in surface circulation. As the large volume and low flow rate of the surface drilling fluid pool, the drilling fluid's rock fragments are separated from the bottom due to gravity settling, while upper drilling fluid re-enters the well for recycling. This process uses drilling fluid recycling technology.

(4) Harmless treatment of drilling sludge

As the petroleum and natural gas industry's pollution source, waste drilling sludge is a polynomial gel suspension system containing clay, weight materials, various chemical treating agents, sewage, sump oil, and drilling cuttings. Its environmental hazards include hydrocarbons, salts, various polymers, lignosulfonates, specific metallic ions (mercury, copper, arsenic, chromium, zinc and lead), and barite impurities. Waste drilling sludge must be treated to prevent environmental pollution.

Waste drilling cutting fluids, sludge, and wastewater are separated from the drilling operation by vibrating screen, desander, and centrifuge, sent to the dispersive cutting and equalizing device by the screw conveyor. Drill cuttings and drilling fluid are the output of drilling operations in the rock mantle, which can be pumped to the debris separator by a hose sorption pump. The filtrate is then sent to the drilling fluid circulator through the pipeline for reuse, while the debris is collected to the brick-making device or landfilled. However, the output of drilling operations in the mud layer is sludge and drilling fluid. They can add gel breaker and coagulant aids through the pump pipeline mixer to the mixing condenser to break the gel and flocculate, separating the aqueous and solid phases in the mud. Then, the primary slurry separation is carried out in a dryer or pressure filter, and the filtrate is sent to the drilling fluid circulator through a pipeline for reuse, while the mud cake is collected for centralized treatment.

Onsite evaluation or supporting materials:

(1) Mining design plan.
(2) Related photos, such as blast hole-defects and filling stations.

19 Mining face quality requirements

Standard score: 30

Rating illustration:

★ Suitable for open-pit mining:

① The working platform is clean, flat, and unobstructed, without sundries and ponding. There is no dangerous rock on the slopes of working or non-working banks. Fifteen points for meeting the requirements

② The falling objects on the non-working banks shall be cleaned up in time with warning lines or safety signs set up at the hidden danger positions. Fifteen points for meeting the requirements

★ Suitable for underground mining:

① The underground mine working face's safety exit is unblocked, meeting the needs of ventilation, transportation, pedestrians, equipment installation, and maintenance with intact support. Fifteen points are rewarded for meeting the requirements
② The working face has no large waterlogging, dross fines, and sundries, with neatly stacked materials. Fifteen points are rewarded for meeting the requirements

★ Suitable for minerals such as petroleum, natural gas, and mineral water:

① Prevent "leakage and venting" without polluting the well site's surface soil by not spilling hazardous chemicals
② Centralized harmless-treat drilling wastes without dropping to land;

Regularly spray water on the bare ground of the well site to reduce dust
Ten points are rewarded for each item that meets the requirements
Evaluation methods: Check the site.

[Interpretation] This article is a quality requirement of working face for different mining of underground mines, open-pit mines, oil, and gas mines.

1. Open-Pit Mining
The working bench angle of open-pit mining affects mine's safe production. However, the working bench angle influences include the ore-rock properties, the method of perforation and blasting, the direction of advance and the ore rock bedding, joints development, and the following contents.

(1) Side uniform: The concave-convex of the face after mining shall not exceed 2 m, while the bag shall not exceed 30 m.
(2) Flat on bottom: Excavate according to the specified elevation, and exhaust the explosion volume. The working berm is smooth, with one point every 30 m, and the height difference does not exceed ± 1 m.
(3) Umbrella eaves: The face's umbrella eaves after excavation shall not protrude 1 m.
(4) The working surface is neat and tidy: large blocks that cannot be loaded are stacked neatly on one side, the working berm has no waterlogging (except for geological reasons), sundries, collapse rocks, and the cables are placed neatly.
(5) Safety retaining wall: The safety retaining wall height of the excavation berm shall not exceed 2 m, and the lower part thickness shall not exceed 5 m.
(6) Final boundary: The excavation is neat and in place, with no over-excavation or under-excavation.
(7) Cable bridge: The placement and angles are safe and reasonable.

(8) Falling objects on non-working banks should be cleaned up in time, with warning lines or safety signs set up at hidden dangers.

2. Underground Mining
Considerations for underground mining face.

(1) The underground mine working face's safety exit is unblocked, meeting the needs of ventilation, transportation, pedestrians, equipment installation, and maintenance with intact support.
(2) The working face has no large waterlogging, dross fines, or sundries.
(3) The lighting of the worksites meets the requirements.
(4) Tools and materials are placed neatly, pipeline hanging is standardized, and the drawing board content is accurate.

3. Petroleum, Natural Gas, and Mineral Water

(1) No spilling of hazardous chemicals to prevent "leakage and venting" and polluting the well site's surface soil.
(2) Real-time disposal of drilling waste, online collection and treatment of fracturing backflow and production without fluids landing.
(3) Regularly spray water on the bare ground of the well site to reduce dust.

Onsite evaluation or supporting materials

Onsite photos of the working face.

2 Mineral Processing

20 Mineral processing and processing technology
Standard score: 60
Rating illustration:

★ Suitable for non-ferrous, metallurgy, gold, non-metal, chemical, coal, and other industries:

① Adopting production equipment and processes with high automation, and low energy consumption and pollutant production;
② Indices of mineral dressing such as recovery rate and concentrate grade meet or exceed the design requirements, making full use of the main metals and associated elements;
③ Choosing high-efficiency and low-toxic agents with little impact on the environment (e.g., the cyaniding chemical room in the gold industry should be isolated and completely enclosed);

④ The content of valuable contents in tailings and waste rock is not higher than the grade that can be processed by existing technology

Fifteen points are deduced for one item that does not meet the requirements until 60 points are deducted

★ Suitable for cement limestone industry:

① The production process is embodied in a short process, low energy consumption and high efficiency;
② The crushing system selects high-efficiency crusher according to the rock breakability;
③ Main production areas such as crushing workshops and conveying corridors are fully enclosed and equipped with dust collection and reduction equipment;

Twenty points are deducted for one failure to meet the requirements

★ Suitable for gravel and architecture stone industries:

① Selecting short-flow, low energy consumption processes and equipment according to the parent rock's material properties, product structure, and capacity requirements with auxiliary facilities that match the production scale and process;
② Dry production is equipped with a dust-cleaning apparatus and runs synchronously with the production equipment. Meanwhile, wet production is equipped with mud powder and water separation, wastewater treatment and recycling systems;
③ The dust-producing point in the production area is closed;
④ When the ground is hardened, the finished sand gravel-aggregate storage yard (warehouse) should be classified or stored separately

Fifteen points will be deducted for one failure to meet the requirements

Basis or standard: *Code for Design of Machine-made Aggregate Plants* (GB51186)

★ Suitable for oil and gas, geothermal, mineral water, and other industries:

① Thirty points are rewarded for selecting moderate crude oil dehydration technical equipment for dehydration, oil–gas separation equipment, and crude oil stabilization technology
② For oil–gas reservoirs associated with carbon dioxide and hydrogen sulfide, and the associated gas content that does not meet comprehensive industrial utilization requirements, 30 points are awarded for effective disposal measures

Evaluation methods: Check data and the site.

[Interpretation] This article is the process requirements for mineral beneficiation or processing, reflecting the environmental protection ability.

Mineral beneficiation and processing focus on cleaner production. Cleaner production adopts improved design, clean energy and raw materials, advanced technology and equipment, improved management, and comprehensive utilization to reduce pollution from the source and improve resource use. Besides, it reduces or avoids production, service, and products used to discharge pollutants, which reduces harm to human health and the environment.

1. Non-Ferrous, Metallurgy, Gold, Non-Metal, Chemistry, Coal, and Other Industries

   Mineral processing is a process of separating useful and gangue minerals from ore by preparation equipment based on physical or physicochemical properties to enrich useful minerals. Mineral resources include metal, non-metal, and burnable organic mineral resources in the industry. Except for a few rich minerals with higher grades, most need to be processed before use, while beneficiation is the primary processing process. Common concepts of mineral processing are as follows.

   Raw ores: The ores have not been processed in the dressing works.

   Concentrates: Products are suitable for industrial applications with improved grade after raw ore beneficiation.

   Tailings: Products whose grades are reduced after raw ore beneficiation cannot be reprocessed with current economic technology.

   Middlings: Products that need to be sorted or treated because the improved raw ore grade after beneficiation does not meet the industrial requirements.

   (1) Purpose and significance of mineral processing
      1) Separating useful and gangue minerals.
      2) Separating useful minerals that are symbiotic with each other and enriching them into separate concentrates.
      3) Eliminating impurities harmful to smelting and other processing processes, improving mineral processing quality.
      4) Comprehensive recovering valuable elements to fully, rationally, and economically utilize mineral resources.

   (2) Mineral properties
      1) Density: The mass of minerals per unit volume is the basis for gravity concentration.
      2) Wettability: The property that minerals can be wetted by water. Minerals that are easily wetted by water are hydrophilic minerals (contact angle with water $<90°$); Otherwise, they are hydrophobic minerals (contact angle with water $>90°$), which are the flotation basis.
      3) Magnetism: Meaning the property of being attracted or repelled by a magnet, which is the basis for magnetic separation.

4) Conductivity: the conductivity of minerals, which is the basis for electrical selection.

(3) Basic operation process of mineral processing.

1) Preparation work: ore crushing and screening, grinding and grading (a few require ore washing and desliming).
2) Grading: gravity, flotation, and magnetic separations.
3) Product processing: concentrate dewatering and tailings storage.

(4) Function of grinding and crushing in mineral processing.

The equipment investment, production cost, power, and steel consumption of crushing and grinding in dressing mill account for the largest proportion: Equipment cost accounts for 60% of that in the mineral processing, while the production cost 40–60%; Meanwhile, electricity consumption accounts for 50–65% of that in the mineral processing, while steel consumption accounts for above 50% of that in the mineral processing. Therefore, the calculation and operation management of crushing and grinding grading determines the dressing mill′s economic benefits.

Two basic procedures of the mineral processing technology: One is dissociation, referring to crushing and grinding large ore to dissociate useful mineral particles; the second is sorting. The dissociated mineral particles are sorted into different products according to their physicochemical properties. Since most useful minerals in nature are in close symbiosis with gangue and embedded in fine particles, full dissociation is a prerequisite for the grading. Crushing and grinding is to dissociate useful minerals and gangues closely connected in the ore. Influencing factors of dissociation include feed size, grinding balls′ number and size gradation, grinding time, and mill type.

Crushing is the process of reducing the particle size of ore blocks, where the dissociation of useful mineral particles occurs. If the crushed product's particle size is not sufficient, and the useful minerals are not sufficiently dissociated from the gangue, the sorting effect is not good; If the particle size is too sufficient with many crushed particles, the useful minerals will be dissociated with a wrong sorting index. Since the particle size of materials that can be processed by grading has a specific lower limit, particles below the limit (over-crushed particles) are difficult to sort. If the crushing process and equipment are not correctly selected with bad production operation management, the final product will be insufficiently dissociated or over-smashed. These two situations will reduce the concentrator's techno-economic indicator. Influencing factors of dissociation include feed size, grinding balls′ number and size gradation, grinding time, and the sander′s parameters and performance.

The root of the concentrator's technical and economic indicators lies in the ore crushing and grinding. Therefore, choosing reasonable crushing and grinding procedures and equipment to reduce the crushing and grinding cost is essential for mineral processing.

(5) Equipment selection
Crushing and screening: Adopting advanced crushing equipment such as large-capacity, high-efficiency ultra-fine crushers equipped with dust separation facilities.
Grinding: Adopting advanced grinding equipment such as barrel mill and high-pressure roller mill with large processing capacity, low energy consumption, and high efficiency.
Classification: Adopting advanced classification equipment such as a high-frequency vibrating fine screener with high classification efficiency.
Grading: Adopting advanced high-recovery, high-automation, large-particle size, medium- and high-intensity magnetic separators and jiggers, vertical ring pulsating high gradient magnetic separators, gas-filled machinery agitator flotation machines, and other separating equipment.
Dewatering filtration: Adopting advanced high-recovery, high-automation high-efficiency thickeners, and large-scale high-efficiency disc filters and other dewatering filtration equipment.
Advanced equipment and technology can achieve clean production with high resource utilization, high recycling, and low waste discharge. Specific requirements are as follows:
1) Adopting production equipment and processes with high automation, low energy consumption, and pollutants.
2) Indices of mineral dressing such as recovery rate and concentrate grade meet or exceed the design requirements, making full use of the main metals and associated elements.
3) Choosing high-efficiency and low-toxic agents with little impact on the environment (for example, the gold industry's cyaniding chemical room should be isolated and completely enclosed).
4) The content of valuable components in tailings and waste rock is not higher than the grade that can be processed by existing technology.

(6) Environmental protection of mineral processing reagent
Environmental protection of mineral processing reagents has no evaluation specifications and standards. Traditional inorganic or organic mineral processing reagents, including collectors, foaming agents, flocculants, activators, depressors, pH regulators, extractants, and diluents, have environmental problems. Environmentally hazardous chemicals include cyanide, hydrogen sulfide, phosphine, fluoride, arsenide, dichromate, copper sulfate, zinc sulfate, carbon disulfide, sulfurous acid, various alkyl xanthates, organic phosphorus, fatty alcohols, aniline, pyridine, hydrocarbon, alkene, ether, ketone, and aldehyde ester. These mineral processing reagents enter the environment through tailings and water bodies and concentrate on endangering natural ecology and human health. Therefore, it is necessary to develop and apply environmentally-friendly mineral processing reagents.

Environmental protection of mineral processing reagents is a relative concept. Non-toxic or low-toxic mineral processing reagents replace high-toxicity reagents, and mineral processing wastewater's recycling are environmental protection improvements. Non-toxic, low-toxic, and environmentally-friendly reagents are pursued by the environmental protection of mineral processing reagents, but their development direction, production methods, and treatment modes lack standards and evaluation standards. The Zhongguancun Green Mine Industry Alliance is committed to forming an environmental protection evaluation system for mineral processing reagents by issuing standards to standardize toxicological indicators, processing modes, and environmental impact assessments.

2. Cement Limestone Industry

Cleaner cement production requires cement enterprises to improve product design, and use clean energy and raw materials (including waste to replace original and fuel materials), advanced production technology and equipment, reducing the electricity, heat, and other unit consumption. The increase in the comprehensive utilization of resources has slowed the exhaustion of limestone, clay, coal, and water resources, reducing cement productions' risks to humans and the environment. Cement production is in harmony with the environment and meets environmental standards.

Cleaner production practices in the cement industry are as follows:

(1) Raw and auxiliary materials and energy

Using a little limestone as cement filler: Cement filler with 3% limestone can reduce cement costs.

Drying the slag with clinker's waste heat: Use the clinker under the kiln to dry the slag at a high temperature. Then reduce the mixed materials′ water content into the grinding, and increase the cement production to achieve comprehensive energy saving.

(2) Technology, process, and transformation

Controlling the raw materials′ water content of the vertical mill: Control the clay's water content, self-dried anthracite, and iron powder to prevent the conveyor belt's bond.

Application of adding compound seed crystal technology: adopt adding-compound seed-crystal calcination technology to improve clinker product and quality.

Gradation optimization of cement grinding chamber diaphragm and "steel without excessive forging": The best location of cement grinding chamber diaphragm and the matching proportion of various steel balls and forgings are found through experiments, improving the production to prevent over grinding.

The comprehensive renovation of kiln-bottom clinker crushing system: The chain scrapper conveyor's seal and the installation of a single-machine bagroom at the jaw break can prevent the material from overflowing during the operation of the chain scraper conveyor and recovering dust during crushing.

Parameter optimization of disc granulator: Adjust the parameters such as the disc granulator speed, the feed opening angle improves the globules quality and reduces the dust moisture content during firing.
Clinker sintering technology optimization: Shallow dark water and gate closure operation process can save energy, realizing shaft kiln calcination reduction.

(3) Equipment maintenance and management
Promotion of bobble calcination technology: As an energy-saving technology in the cement industry, bobble calcination adjusts the disc granulator's operating parameters, such as angle, rotation speed, and water addition. The pelletization grain diameter is about 5 mm, saving energy and improving the specific surface area and clinker quality.
Modification of the packers′ bagroom: Install "rag dolls" on the top of the raw material, clinker, and cement warehouses to avoid the dust lost from the warehouse top.
Transformation of clinker storage blanking: Installing a hose in the clinker, slag, and gypsum warehouse to implement a soft blanking landing and reduce the dust caused by the drop.
Interface seal of cement grinding head bucket-elevator discharge pipe and grinding head warehouse: the equipment room is sealed to avoid dust.

(4) Waste recycling
Reclamation of the cement grinding mill screen tailings: the cement grinding mill screen tailings are discarded as waste in the past, but now it is sorted to be a raw material for cement grinding mill operations.
Reusing the cooling water of cement grinding mills: Lead the cooling water of two cement grinding mills into the recycling reservoir.
Adopting cleaner production measures to achieve energy saving, noise, and dust reduction. Requirements for evaluation indicators are as follows:
1) The production process is embodied in a short process, low energy consumption, and high efficiency.
2) The crushing system selects a high-efficiency crusher according to the rock breakability.
3) Main production areas such as crushing workshops and conveying corridors are fully enclosed and equipped with dust collection and reduction equipment.

3. Gravel and Architecture Stone Industry
The industry uses construction waste and stone materials to produce sand gravel aggregates and integrates pebble mechanism sand and gravel river dredging. The increased proportion of industrial solid wastes recycling is conducive to the restoration of abandoned quarries and a circular economy. Extensive use of stone and mud powder in sand gravel production produces cement, wall materials, mortar, environmentally-friendly permeable bricks, and other products. The realization of recycling and comprehensive reuse of wastewater and solid wastes

in sand gravel enterprises improves resource utilization and added value. The following should be considered in the production process:

(1) Select short-flow, low energy consumption processes, and equipment based on the parent rocks′ material properties, product structure, capacity requirements with auxiliary facilities that match the production scale and process.
(2) Dry production is equipped with dust-cleaning apparatus and runs synchronously with the production equipment. Meanwhile, wet production is equipped with mud powder and water separation, wastewater treatment, and recycling systems.
(3) The dust-producing point in the production area is closed.
(4) When the ground is hardened, the gravel aggregate finished product storage yard (warehouse) should be classified or stored separately.

4. Oil–gas, Geothermal Mineral Water, and Other Industries

Crude oil pretreatment is the desalting and dewatering process. Oil entrains oilfield associated saline and solid sediments in the recovery process, causing the crude oil to contain water and salt. The water and salt contents of crude oil are related to the oilfield, the limit of years, and the enhanced production methods (such as water and gas injections). The water content can reach several percent, ten percent, or even tens of percent, while the salt content can reach tens to several thousand or even tens of thousands of milligrams per liter. As crude oil becomes heavier and deteriorates, water and salt content are also increasing.

Impurities of crude oil include:

Bits of solid impurities such as sediment and rust;

Water in oil (associated water in the oilfield, water injection during exploitation);

Crude oil contains salt (e.g., sodium, potassium, calcium, and magnesium chlorides. Content: sodium salt higher than calcium salt, few magnesium and potassium salts; and sulfate, carbonate, and oil-soluble organic acid salts).

Oilfield first adopts the sedimentation to remove part of the water, and solid impurities (silt and solid salt), and the water content of the exported crude oil is controlled to be less than 0.5% and the salt content less than 50 mg/L. Specific requirements are as follows:

(1) Thirty points will be obtained to select moderate crude oil dehydration technical equipment for dehydration, appropriate oil–gas separation equipment, and crude oil stabilization technology.
(2) For oil–gas reservoirs associated with carbon dioxide and hydrogen sulfide and the associated gas content that does not meet comprehensive industrial utilization requirements, 30 points are awarded for effective disposal measures.

Onsite evaluation or supporting materials:

(1) Design plan for beneficiation or mineral processing.
(2) Check onsite mineral processing equipment.
(3) Photos of equipment and process.

3 Mine Environmental Restoration and Control and Land Reclamation

21 Scope requirements

Standard score:30

Rating illustration: According to the mine geo-environment restoration and control and land reclamation plan, the designated areas shall be treated and reclaimed, such as dumps, open-pits, special mining roads, industrial mining squares, subsidence areas, gangue dumps, and contaminated mine sites. Five points will be deducted for each area that should have been treated and reclaimed but fail to follow the plan in time

Evaluation methods: Check data and the site

Basis or Standard: *Mining Geo-Environmental Protection and Land Reclamation Plan*

[Interpretation] This article requires the mine environmental restoration and control and land reclamation scopes.

Mine geo-environment refers to the objective geo-object that interacts with the lithosphere, hydrosphere, and biosphere affected by mining activities. Mine geo-environmental problems refer to the geo-environmental destruction phenomena caused by mining activities, including surface subsidence, ground fissures, collapses, landslides, mudslides, aquifer destruction, and topography and landscape devastation in mining areas.

Land reclamation refers to comprehensively rectifying the damaged land to restore it to an available state.

1. Negative Effects of Resource Development on the Environment

   The environmental effect refers to how natural processes or human production and living activities will pollute the environment, causing changes in the environmental system′s structure and function. The resource development process has positive and negative effects on the environment, while this book only introduces the adverse effects.

   (1) Types of adverse effects
      1) The topography and landform changes, and plants′ soil and water conditions are affected.

2) Land subsidence and surface cultivated land's paludification.
3) Ground collapse, rivers cut-off, and ground building foundation destruction.
4) The land occupation and dust pollution of tailings ponds and waste rock hills.
5) Soil desertification and soil and water loss.
6) Slope landslides and debris flows in open-pit mines and dumps.
7) Corrosion of mine acid wastewater to soil and foundation water softening.

(2) Prevention measures

1) Adopting open-pit mining, designing open-pit side slope step surface, and pre-considering technical measures for slope stabilization, such as soil anchoring and planting.
2) Underground mining fully predicts the damage to the surface caused by movement, subsidence, landslides, and uses filling mining.
3) Glissando is adopted for mine blastings, such as setting up shock-absorbing bands and wave-blocking walls.
4) Once the stratum subsidence is found in mining, the bed-separation grouting shut-off technique can be used to inject the cementing material into the stratum from the ground grouting hole. The filling and cementation of the broken abscission zone can prevent ground settlement.

2. Implementing the Principle of Integrating "mining, Governing and Restoring"

According to the principle of integrating "mining, governing and restoring," the mines' geo-environment shall be restored promptly, and land occupied and damaged by mines shall be reclaimed.

The principle of integrating "mining, governing and restoring" is a mine governance model that combines "planning, mining, and governance." This principle turns ecological governance from passive to active and from restorative to protective. Protective governance combined with pre-planned governance achieves the unification of ecological and economic benefits and embodies the guiding ideology of "protecting in development and developing in protection".

(1) The coordinated development of exploitation and protection refers to environmental protection before, during, and after exploitation. Mining areas and the surrounding environment, especially the geological conditions, shall be explored following laws and regulations before the development. In the development, the cultivated land, farmland, and vegetation area will be less or not occupied, and the waste generated will be treated by the pollution discharge standards. After development, measures should be taken to restore the mining area's surrounding environment, especially the destroyed land resources and vegetation. Land reclamation measures such as simultaneous stripping and reclamation can reduce costs and restore greenery.

(2) Mining enterprises prospecting consider the *Mine Geo-environment Protection and Land Reclamation Plan* in advance. Make overall arrangements and implement the plan during mine design, production, and shutting.
(3) The mine geo-environment governing efficiency and land reclamation rate meet the Mining Geo-environmental Protection and Land Reclamation Plan requirements on time.

3. Policy Requirements
The *Notice on Achieving the Formulation and Submission of Mine Geo-environmental Protection and Land Reclamation Plans* (Land and Resources Planning [2016] No. 21) requires mining enterprises to formulate and submit *Mine Geo-environmental Protection and Governance Restoration Plans* and *Land Reclamation Plans* together. The plan should clarify the purpose, requirements, ways, and means, and formulate an implementation plan for mines′ geo-environmental comprehensive management to realize the mine subsidence range's management and control.
Contents of the *Mining Geo-environmental Protection and Land Reclamation Plan* are as follows:

(1) Basic situation of the mine.
(2) Basic information of the mining area.
(3) Assessment of mine geo-environmental impact and land damage.
(4) Feasibility analysis of mine geo-environment management and land reclamation.
(5) Mine geo-environment management and land reclamation projects.
(6) Mine geo-environment management and land reclamation work deployment.
(7) Estimated expenditure and schedule.
(8) Guarantee measures and benefit analysis.
(9) Conclusions and suggestions.

4. Scope of Land Reclamation
Land to be reclaimed includes but is not limited to:

(1) Land damaged by surface excavation, such as open-pit mining, brick burning, sand excavation.
(2) Land where the surface collapses for underground mining.
(3) The land occupied by solid waste such as mining stripped materials, waste rocks, slag.
(4) Land occupied by dumps, special roads in mining areas, and mine industrial sites.
(5) Reclamation or green open-pit′s final platform in time.
(6) Reclamation and green of areas that affect open-pit mine dumps′ environment and other mining areas.

Onsite evaluation or supporting materials:

(1) *Mining Geo-environmental Protection and Land Reclamation Plan.*
(2) Onsite photos.

22 Governance requirements
Standard score: Ten
Rating illustration:

① Recovering various sites after management, which are coordinated with the surrounding natural environment with landscape effects
② If the exposed area caused by open-pit mining has a greater impact on the surrounding landscape, measures to reduce the adverse impact shall be taken;
③ Open-pit mining shall meet the requirements for the open-pit mining platforms setting up and the reclamation and afforestation

Four points will be deducted for one item not meeting the requirements until ten points are deducted
Evaluation methods: Check data and the site
Basis or standard: *Mine Geo-environmental Protection and Land Reclamation Plan, Land Reclamation Quality Control Standard* (TDT1036), and other documentary proof materials

[Interpretation] This article is a requirement for geo-environment governance and land reclamation.

Mineral reserves′ development has accumulated large waste rocks and tailings in the mining area. The industrial wastewater produced contains large suspended solids, which destroys the mining area's ecological environment. It is manifested in the annual reduction of vegetation coverage, river channel sedimentation, water acidification, soil erosion, wind and sand damage, and severe soil erosion.

1. Water and Soil Conservation Requirements
   Erosion control degree of soil erosion: The percentage of the area that meets the standards for soil erosion harness within the limitation of liability for preventing and controlling soil erosion in the project to the total area of soil erosion.
   Soil loss control ratio: The ratio of allowable soil loss within the limitation of liability for preventing and controlling soil erosion in the project to the annual average soil loss per square kilometer after treatment.
   Muck protection rate: The percentage of permanently wasted slag and temporarily piled solids protected within the limitation of liability for preventing and controlling soil erosion in the project to the total number of permanent waste slag and temporary piled soil.

Topsoil protection rate refers to the percentage of the protected topsoil amount within the limitation of liability for preventing and controlling soil erosion in the project to the strippable topsoil′s total amount.
Forest-grass vegetation restoration rate is the percentage of forest-grass vegetation area within the limitation of liability for preventing and controlling soil erosion of the project to recoverable forest-grass vegetation area.
Forest-grass coverage: The percentage of forest-grass vegetation area in the total area within the limitation of liability for soil erosion prevention and control of the project.
The *Standards for the Prevention and Control of soil erosion in Production and Construction Projects* (GB/T 50434-2018) clarifies the standards for preventing and controlling soil erosion in production and construction projects.

(1) Standard grades of soil erosion prevention and control for production and construction projects

    The soil erosion control standards for production and construction projects are divided into levels one, two, and three.

    1) The following projects implement level-one standards: Key prevention and control areas for soil erosion determined by the People's Governments at all levels and relevant institutions, drinking water source protection areas, protection and reserved areas of the level-one water function areas, nature reserves, places of world cultural and natural heritage, scenic spots, geological parks, forest parks, unavoidable and important wetlands, and urban areas at the county level and above;
    2) Projects locate around lakes and built reservoirs, within 3 km of confluence on both banks of rivers above Grade 4, or projects with towns and residential areas within 500 m, and not in the level-one standard areas, implementing the level-two standard;
    3) Projects, locating outside levels-one-and-two standard areas, implement the third level standards.

(2) Soil and water conservation areas

    Soil and water conservation areas include northeast black soil area, northern windy desert area, northern earth-rock mountain area, northwest loess plateau area, southern red soil area, southwest purple soil area, southwest karst area, and Qinghai-Tibet plateau area. For indicators, see *Standards for Preventing and Controlling Soil Erosion in Production and Construction Projects* (GB/T 50434-2018).

(3) Standard indicators for soil erosion prevention and control

    The standard indicators for soil erosion prevention and control include soil erosion control degree, soil erosion control ratio, muck and topsoil protection ratios, forest-grass vegetation restoration rate, and forest-grass coverage rate.

1) The control index value of the newly added disturbance band during the production period shall not be lower than that during the construction period, while the annual index value for other areas shall not be lower than the design level.
2) The same project involves more than two prevention and control standard grade areas to determine the corresponding segment's index value.
3) When calculating the values of various prevention and control indexes for mining and water engineering projects, the open-pit mining area and a water area of water engineering shall be deducted from the prevention and control responsibility′s area. Meanwhile, when the cultivated land area is restored, forest-grass coverage shall be deducted from the prevention and control responsibility′s area.
4) Soil erosion control degree, forest-grass vegetation restoration rate, and coverage rate can be adjusted according to the degree of drought according to the following principles: ① There is no quantitative requirement for forest-grass vegetation restoration rate and coverage rate in extremely arid areas, while 5–8% can reduce soil erosion control degree. ② Soil erosion control degree, forest-grass vegetation restoration rate, and coverage rate in arid areas can be reduced by 3–5%.
5) The soil loss control ratio should not be less than 1 in areas with mild erosion and reduced by 0.1 to 0.2 in moderate or above erosion areas.
6) The muck protection rate of the Zhongshan District project can be reduced by 1 to 3%, and that of the project in extremely high-mountain areas can be reduced by 3 to 5%.

(4) Other requirements

1) The soil and water conservation plan of the mining enterprise implements the scope and responsibility of the construction party's soil erosion prevention and control according to the principle of "whoever develops protects; who causes soil erosion is responsible for governance."
2) Soil and water conservation project types: Construction projects change small terrain, intercept surface runoff, and increase soil infiltration (increasing soil infiltration rate).

   Soil and water conservation measures include engineering measures, plant measures, and temporary measures. The former includes artificial soil covering, manual excavation of drainage ditches, and installing mortar retaining walls and slope cuttings, and plant measures include cave-like soil preparation and plant setting. The latter includes the protection project of straw-bags filled with soils.

2. Coordinating with the Surrounding Natural Landscape

   (1) Land reclamation shall be carried out by the requirements of eco-environment protection and restoration management of the damaged land and land quality requirements after restoration.
   (2) Coordinating with the plant community in the mining areas′ ecological zone. The greening species are interspersed and connected, with the basic requirements of living long and prospering, to improve the eco-environment and aesthetic.
   (3) Land reclamation determines the proportion and configuration of various plants according to the enterprises′ types and their production characteristics, pollution nature and degree, local natural conditions, surrounding environmental conditions, and the reclamation effect to be achieved.
   (4) Landscape design and environmental beautification should be carried out according to local conditions (not required for geo-environmental governance).
   (5) There are no gullies or large soil areas directly exposed to the mining areas and waste dumps.
   (6) The site's use types to be reclaimed: Coordinate with the local topography, landform, and environment.
   (7) The topography after restoration and control is coordinated with the surrounding environment and surface water system in a small area. The topographic slope is determined by water and soil conservation requirements and rock mass stability, conforms to the soil stable slope angle of 20–40°, and the stable slope angle of 30–50°.

3. Open-Pit Mining Reduces the Adverse Impact on the Visible Landscape

   (1) Mining should determine the mining face′s advance direction and adopt measures such as delaying external mountain mining, human-made landscape covering the mountain trauma, and concave mining to reduce the adverse impact on visible landscapes.
   (2) Mining is integrated with surrounding buildings, environments, cultures, and landscapes.

4. Setting up the Open-Pit Mining Terminal Platform
   The steps used for mining are the working steps, whose flat plates are the working ones. On the contrary, they are non-working steps, and flat plates are platforms, which can be divided into safety, sweeping, or transportation ones depending on their uses.
   The security platform blocks the falling rocks, slows the final slope angle, ensuring the final pit slope's stability (See Fig. 1) and the lower level′s safety. The width is narrow, within 1/3 of the mining level height.
   The sweeping platform can block the falling rocks and clean them with sweeping equipment (excavator or bulldozer) to ensure safety. Generally, a cleaning platform is set on the four slopes every 2 to 3 steps, the width of which is determined by the cleaning equipment.

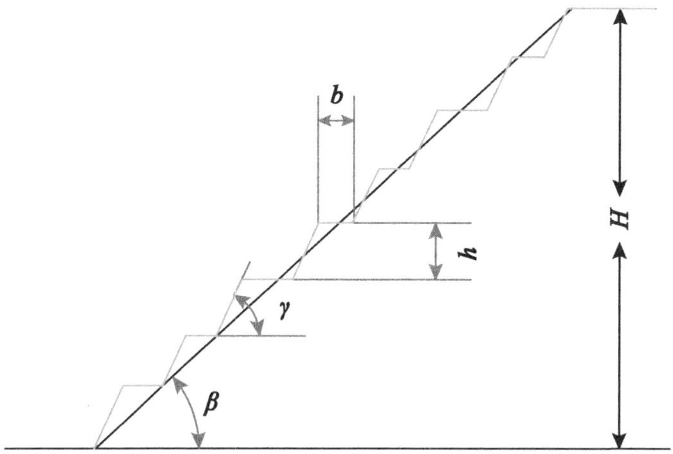

Fig. 1 Elements of the final pit slope. *Note* H is the stope height; h the mining level height; β the stope side angle; γ the bench slope angle; b the safety platform width

The transportation platform is the transportation corridor between the working banks and the main access. Suppose on the main access's non-working and the end slopes, its width is determined by the used transportation modes and lines' number.

(1) Height of production steps.
According to the *Safety Regulations for Metal and Non-metal Mines* (GB 16423-2006), the height of production steps should meet the requirements of Table 1.
If the step height after the mining bench combination exceeds the regulations in Table 1, it shall be subject to technical demonstration, safety, and design.
The blasting muck pile height shoveled by the excavator or loader should not be greater than 1.5 times the maximum mechanical digging height.

Table 1 Requirements for production steps' height

Ore properties	Mining operation modes		Bench height
Soft rock soil		Not blasting	Not more than the mechanical digging height
Hard and stable rock	Mechanical shovel loading	Blasting	Not more than 1.5 times the maximum mechanical digging height
Sandy ore-rock			Not more than 1.8 m
Soft rock	Man-made recovery		Not more than 3.0 m
Hard and stable rock			Not more than 6.0 m

The mining level height of small mines designed and produced in China for open-pit mines is 8–10 m, while large and medium-sized mines are 10–12 m. During using excavators with a bucket capacity of 8 m^3 or more for a large open-pit mine, the mining level height can reach 12–15 m.

1) To ensure the slope stability, safety, and the sweeping platforms must be retained after the level mining. The safety platform width at each level is not less than 2 to 3 m, and the sweeping platform width is not less than 4 to 5 m every 2 to 3 levels.
2) Open-pit mines' location has poor engineering-geological conditions and stubbornness, old rocks, and incomplete structure with weak geological structures and unstable rocks, which are easy to cause landslides. Too steep or high slope angle will cause local instability. Therefore, high and steep slopes must be mined from top to bottom in benching or slicing according to the design. The slope angle is not more than 55°, while the final slope is not more than 60°. Besides, the layered rock drilling platform's width is not less than 4 m, while the stripping face of the quarry that needs stripping is 4 m above the mining face.
3) The terminal platform carries out land reclamation following the plan of the land reclamation plan.

5. Requirements for Reclamation and Greening
   The premise of open-pit mining's reclamation and greening is that the slope and the original engineering facilities are stable.

Onsite evaluation or supporting materials:

(1) *Mine Geo-environmental Protection and Land Reclamation Plan.*
(2) Onsite photos.

23 Land reclamation function requirements

Standard score: Ten

Rating illustration: The treated sites should restore the land's basic functions and realize sustainable land use according to local conditions. Ten points for meeting the requirements

Evaluation methods: Check data and the site

Basis or standard: *Mine Geo-environmental Protection and Land Reclamation Plan, Land Reclamation Quality Control Standard* (TDT1036), and other documentary proof materials

[Interpretation] This article requires that the sites be used for sustainable land use after treatment.

1. Functions of Land
   The first is the bearing function. Land's physical properties carry everything, including biotic and abiotic, and become their homes. Land's bearing function allows living things such as animals and plants, and other non-living things such as buildings, structures, and roads on the earth. Without land, everything has no place. As the ancients said, if there is no skin, there will be no hair. Land's bearing function makes it a place and space for human life and production activities.
   The second is the nurturing (production) function. "The earth is the origin of all things, and the root of all life." Within a certain depth and height of the land, the nourishing (production) capacity breeds everything. For example, the soil contains nutrients, water, and air, and receives the light and heat from the sun, which are the basic conditions for the growth and reproduction of organisms on the earth. Land's nourishing function is embodied in primary and secondary production, providing the necessary agricultural and livestock products for human survival.
   The third is the storage (resources) function. Land can be regarded as a human warehouse, with rich mineral resources (such as gold, silver, copper, and iron), energy resources (such as oils, coals, water power, and natural gases), and building materials (such as sands, stones, and soils).
   The fourth is the landscape function. From the perspective of landscape architecture, the landscape is the mosaic of the ground ecosystem and the carrier of the natural and cultural ecosystem. Various landscapes naturally formed by the land, including beautiful mountains, vast seas, swift currents, cascading waterfalls, and endless fertile fields, provide humans with abundant scenery resources.
   The fifth is the saving and value-added (reproduction) function. As an asset, land price is rising with the continuous expansion of people's demand for land. Therefore, land investment can achieve the effect of saving and value-added.

2. Land Ecosystem
   Lands' basic function refers to its ecological function, and the core is the ecological system. The land ecosystem is an organic whole composed of ecological factors such as land, natural environment, technology, policies, and people. Any factor's change in the system will break the land's original ecological balance in nature. Although the land ecosystem has a specific restoration function, it is limited and cannot be restored beyond the limit.
   Understanding the land ecological functions from different levels and aspects:

   (1) Understanding the interaction between organisms and the land environment from the land ecosystem inside, including the requirements and adaptability of organisms to land environmental conditions and the land environment transformation by organisms.

(2) Land ecosystem's general awareness of a land unit, such as the forest ecosystem's ecological functions such as water conservation, wind sheltering and sand fixation, soil and water conservation, climate regulation, and farmland protection.
(3) Besides the local land unit, a unit's land ecological effect also affects the nearby areas, such as the upstream and downstream areas that are harmed by water and soil conservation, wind, and sand, even expanding to form a regional effect.
(4) To understand the environmental effect of land ecological effect, we should see the greater economic effect, such as environmental value.

3. Land Resources' Sustainable Use
Land resources' sustainable use refers to the reasonable development, use, management, and protection of land resources under specific time and regional conditions. The coordination of the relationships among man, land, resources, and environments should meet current and future generations' survival and development.

Onsite evaluation or supporting materials:

(1) There are plans for the use or planning of reclamation sites, such as forest land, cultivated land, and industrial land.
(2) Without any planning and utilization, only restoring the ecosystem will not score.

24 Ecological function requirements

Standard score: Ten

Rating illustration: The treated site should meet the following: ① The region's overall ecological function is protected and restored. ② There is no threat to animals and plants. Five points are deducted for one item not meeting the requirements

Evaluation methods: Check data and the site

Basis or standard: *Mine Geo-environmental Protection and Land Reclamation Plan, Land Reclamation Quality Control Standard* (TDT1036), and other documentary proof materials

[Interpretation] This article is a requirement for ecological functions.

The geo-environment is secure and stable without polluting the surrounding environment. Besides, the coordination with the surrounding natural environment and landscape can protect and restore the area's overall ecological function.

The protection and restoration of the region's overall ecological function refer to the realization of natural growth without artificial maintenance and the coordination and unification of the surrounding ecosystems.

Reclamation sites meet the following requirements.

1. Reclamation Site Requirements
   Reclamation utilizes the land occupied and damaged by the mine, while the reclamation quality complies with the *Land Reclamation Quality Control Standard* (TD/T 1036—2013).

   (1) The covering material used as the reclamation site is non-toxic. The toxic and harmful components in the reclamation site should be disposed firstly, and an isolation layer should be installed according to the waste nature and site conditions before covering.
   (2) The covered reclamation site shall be normative and smooth. The bulk density of the overburden meets the reclamation and utilization requirements.
   (3) Drainage facilities of the reclamation sites meet the requirements, while the flood control standards meet the local requirements.
   (4) Reclamation sites have pollution control measures, including air, surface water, and groundwater.
   (5) Roads and traffic arteries on the reclamation site layout reasonable.

2. No Threat to Animals and Plants

   (1) Choose local wild species, avoiding introducing alien species to threaten local species.
   (2) Rationally match species to form a stable symbiotic phytocoenosium and reconstruct the ecosystem.
   (3) Set guardrails, purse seines, and other facilities in unsafe areas to protect animals′ lives.
   (4) Construct forest fireproofing belt if necessary.
   (5) No grazing.
   (6) Setting signs such as "Dangerous Here."

Onsite evaluation or supporting materials:

(1) Coordinating with the surrounding ecological environment without affecting or damaging the surrounding ecological system.
(2) There are protection measures for animals and plants, while the reclamation site meets clean and pollution-free requirements.
(3) Checking relevant data or *Mining Geo-environmental Protection and Land Reclamation Plan.*

4 Environmental Management and Monitoring

25 Environmental protection facilities

Standard score: Six

Rating illustration:

① Four points are rewarded for complete environmental protection facilities and effective operation;

② Two points are rewarded for effective operation

Evaluation methods: Check data and the site

Basis or Standard: Acceptance data of environmental protection facilities

[Interpretation] This article is a requirement for environmental protection facilities.

Environmental protection facilities are the equipment and devices required to control the substances that affect the environment in the production and operation of industries, commerce and service industries to meet legal requirements, as well as environmental monitoring equipment.

1. Environmental Protection Facilities' Types
   Environmental protection facilities include public environmental sanitation facilities, filtration and dust removal equipment, sewage treatment equipment, air purification equipment, solid wastes treatment equipment, noise prevention equipment, environmental monitoring equipment, disinfectant and preservative equipment, energy-saving equipment, environmental sanitary equipment, environmental protection materials and reagents, and environmental protection apparatus.

2. Environmental Protection Facilities' Acceptance
   Environmental protection facilities shall pass the acceptances of the environmental protection or the environmental protection facilities′ operation effect involved in the project closeout, and ensure effective operation through maintenance.

   The *Regulations on Environmental Protection Management of Construction Projects* have the following provisions: After the construction project is completed, construction units shall apply to the environmental protection administrative department that examines and approves the report, report table or registration form of the project′s environmental impact for the completion and acceptance of the environmental protection facilities required for the project. For construction projects that require trial production, construction units shall, within three months from the date the construction project is put into trial production, apply for the completion and acceptance of the environmental protection facilities required for the project.

For projects that have passed the environmental protection facilities′ acceptance, the following should be noted during evaluating green mines:

(1) Acceptance opinions of the environmental protection department consistent with the scale.
(2) Environmental protection status assessment and filing of construction projects (screenshot of competent department website).

3. Management and Maintenance

(1) Check environmental protection indicators' completion and environmental protection facilities' operation and maintenance, adjust the equipment's operating conditions, and eliminate defects to ensure environmental protection facilities' normal and effective operations.
(2) Strictly implement defects management regulations to ensure that environmental protection facilities are operating in the best conditions, and promptly notify the maintenance units to deal with defects.
(3) Record, count, analyze, summarize, and report materials and statements as required, responsible for their correctness.

Onsite evaluation or supporting materials:

(1) Acceptance certificate of environmental protection facilities.
(2) Operational ledger of environmental protection facilities.
(3) Two screenshots on the Internet after the enterprises' independent environmental protection acceptance (corporate website, local website, or the Ministry of Environmental Protection's unique website).

26 Environmental management system certification

Standard score: Four

Rating illustration:

Four points are rewarded for obtaining the environmental management system certification

Evaluation methods: Check certificate

Basis or standard: ISO environmental management system certification

[Interpretation] This article encourages mines to obtain the environmental management system certification.

EMS belongs to the management system of enterprises or other organizations. It is used to formulate and implement environmental policies and manage environmental factors, including the organizational structures, planning activities, responsibilities, conventions, procedures, processes, and resources required to formulate, implement, realize, assess, and maintain environmental policies.

China's new version of the EMS certification standards is *EMS Requirements and Guidelines* (GB/T 24001-2016), equivalent to *EMS-Requirements and Guidelines* (ISO14001: 2015).

Onsite evaluation or supporting materials:

Environmental management system certification.

27 Environmental monitoring system

Standard score: Five

Rating illustration:

Five points are rewarded for establishing a long-term environmental monitoring mechanism and an environmental monitoring system

Evaluation methods: Check data

Basis or standard: Environmental monitoring system

[Interpretation] This article requires mining areas to establish environmental mechanisms and environmental monitoring systems.

Refer to the monitoring requirements in "Chapter 6 Solid Wastes Stacking".

The environmental monitoring system is applicable to monitoring and managing the pollution sources within enterprises, which stipulates the environmental monitoring tasks, control indicators, and methods. It includes monitoring and supervising the "three wastes" (containing noise) pollution generated during the production process; sorting out and reporting the monitoring data to the relevant management department to establish a monitoring ledger; completing other temporary monitoring tasks in time.

To monitor changes in the mine environment in real-time and ensure that the environment meets the geo-environmental protection requirements, mining enterprises should establish a real-time environmental monitoring mechanism and allocate full-time management and monitoring personnel. Specific requirements are as follows:

(1) Establish and improve a long-term monitoring mechanism during and after mining to monitor land damage, reclamation's stability, and quality dynamically.
(2) According to the mines' mining characteristics, establish a monitoring mechanism for the whole process with emergency plans.
(3) Monitor mines' slope stress to prevent geological disasters.
(4) Dynamically monitor groundwater, surface water, soil environment, surface deformation, and geological disasters.

(5) The stripping of surface soil and the discharge of waste residues from open-pit mining shall comply with safety, environmental protection, and monitoring regulations.
(6) Dynamically monitor and timely treat the land that is sinking and difficult to manage.
(7) Dynamically monitor pollutants such as domestic wastewater, noise, dust, and other pollutions, and prepare emergency plans for environmental protection disposal.
(8) Self-monitor the soil environment in the mining area every year and publish the results.

Onsite evaluation or supporting materials:

Environmental monitoring system.

28 Environmental monitoring equipment

Standard score: Five

Rating illustration:

Five points are rewarded for installing automatic monitoring and electronic display equipment of noise and air pollutants in mining areas

Evaluation methods: Check the site

[Interpretation] This article requires the mining area to be equipped with environmental monitoring equipment.

This article requires mining enterprises to set up small-environment climate stations in locations with large turnover or noise and dust emissions in the mining area. Environmental monitoring objects are PM_{10}, $PM_{2.5}$, noise, temperature and humidity, wind power, wind speed, wind direction, atmospheric pressure, illuminance, oily smoke, carbon dioxide, oxygen, and ammonia in the air. The small-environment climate monitoring station's function is that the employees inside the mining area and external visitors can understand the dust, noise, humidity, and temperature changes in the mining area and formulate measures to improve and optimize the environment.

1. Environmental Equipment

   Environmental monitoring equipment is divided into the following categories by the function, use environment, industry, and nature:

   (1) Special monitoring instruments for air quality and pollution source exhaust gas.
   (2) Special monitoring instruments for water quality and produced water.
   (3) Portable on-site emergency monitoring equipment.
   (4) Other element monitoring instruments.
   (5) Laboratory general analytical instruments and equipment.

Due to the equipment's different measurement factors such as functions, principles, technologies, costs, and application scenarios, the same parameters differ in measurement range and accuracy.

2. Continuous Automatic Monitoring System

   (1) Noise
   The continuous automatic noise monitoring system includes a noise monitoring terminal, all-weather outdoor microphone units, fixed stations, wide-band noise measurement, spectral noise measurement, and LED display.

   (2) Air pollution
   As a network composed of monitoring instruments, data communications, and computers, the continuous automatic monitoring system for air pollution includes several monitoring sub-stations, one monitoring center, and one data communication system.

Onsite evaluation or supporting materials:

Onsite photos taken by the small-environment climate monitoring station.

29 Emergency response mechanism

Standard score: Five

Rating illustration:

Five points are rewarded for establishing an emergency response mechanism and emergency response measures for emergency environmental events

Evaluation methods: Check data

Basis or standard: Emergency response system

[Interpretation] This article requires establishing an environmental emergency response mechanism.

Enterprises shall establish an emergency response mechanism and maintain operating procedures to determine potential accidents or emergencies. Emergency preparedness and response can minimize the possible accident consequences and prevent or reduce the possible accompanying green mines impacts.

Mining enterprises should compile environmental emergency response systems, including environmental emergency response and plans. Emergency management system refers to the following content.

1. Emergency Management Principles

   (1) Adhere to people-oriented and prevention first. Monitor and supervise environmental factors by implementing supervision and management. Meanwhile, establish an environmental factor anti-risk system to improve

the ability to prevent and deal with sudden environmental pollution incidents.
(2) Strengthen coordination and cooperation among various departments with unified leadership to improve quick-reaction capabilities.
(3) Make mental, material, technical, and work preparations for sudden environmental pollution events with existing resources. Besides, ensure that the emergency response system is always ready to respond quickly and effectively by intensive training and drills.

2. Emergency Management Agency Establishment
The emergency command center is the highest leading organization for emergency management of environmental emergencies. Emergency organizations consist of command management, rescue teams, technical support, and related guarantees.

3. Operating Mechanism

(1) Identifying and evaluating environmental factors can be carried out to discover, report, and dispose environmental hazards in time.
(2) Emergency incidents of environmental pollution or ecological (impact) damage occurring within the enterprise should be reported promptly, regardless of the incident′ cause and influence extent, as well as the accident rank.
(3) After an emergency occurs, the onsite and emergency personnel at the incident source shall report the environmental incident and initiate an emergency response plan according to their responsibilities and prescribed authority. Timely and effective pre-disposal can control the situation spread.
(4) Information release of environmental emergencies should be timely, accurate, objective, and comprehensive. The occurrence of major environmental accidents should be reported to the local environmental protection department in time, and follow-up reports should be done according to the accident disposition. Besides, enterprises should release brief information and countermeasures to employees.

4. Emergency Protection
According to the responsibilities′ division, all departments are required to plan the workforce, material resources, financial resources, transportation, and communication guarantee for responding to environmental emergencies, ensuring the smooth progress of emergency rescue work.

5. Emergency Training and Drill

(1) Emergency rescue personnel are composed of young and skilled vital employees who participate in business knowledge training every year.
(2) Organizing emergency rescue teams and front-line employees to conduct a drill for possible incidents every year.

(3) After the drill, the emergency drill team summarizes and evaluates the drill results promptly manner, implements improvement measures to improve the plan.

Onsite evaluation or supporting materials:

Environmental emergency response mechanism.

30 Dynamically monitoring the mines geo-environment
Standard score: Five
Rating illustration:
Five points are rewarded for dynamic monitoring of mine geo-environment, such as surface deformation
Evaluation methods: Check data and the site
Basis or standard: Dynamic monitoring records

[Interpretation] This article requires monitoring the mines geo-environment dynamically.

Mine geo-environment is a relatively independent environmental system composed of the lithosphere (rock, ore, soil, groundwater, and geological processes and phenomena) affected by mining activities in the ore deposit and its surrounding areas, as well as its connection with atmosphere, water and biosphere (material exchange) and energy flow.

Surface deformation is a destruction phenomenon that deforms the surface configuration due to internal and external dynamic geological action and human activities. Besides, surface deformation that causes economic losses and casualties is surface deformation geological disasters.

Mine surface deformation monitoring includes slope and land collapse monitoring.

1. Monitoring Content of Geological Disasters
   Monitoring of land subsidence (collapse) in the mined-out area: collapse pits′ number and the maximum depth, collapse area, waterlogging depth, collapse destructiveness, and subsidence value.
   Monitoring of geological disasters such as mountain cracks, landslides, collapses, and mudslides: The frequencies this year, the harm caused, the number of geological disasters′ hidden trouble sites or areas, and the number of hidden trouble sites or areas that have been treated.

2. Monitoring Method of Geological Deformation

   (1) Land collapse monitoring in mined-out areas: Remote sensing monitoring technology is used for mining areas with large collapse areas; high-precision GPS, borehole inclinometers, and total stations are used

to monitor key mining areas; manual field surveys and measurements are used for others. Monitoring collapsed cracks′ length, visible depth, width, and others.

(2) Surface subsidence monitoring in mining areas: The key mines are automatically monitored by on-site buried bedrock marks, and the others by high-precision GPS.

(3) Slope monitoring items include inspection and patrol, deformation monitor, stress monitor, vibration monitor, and hydrological monitor. Inspection and patrol require manual labor, while other items can be automatically monitored.

(4) Mountain cracking monitoring in mining area: Manual field investigation and measurement are adopted.

Onsite evaluation or supporting materials:

(1) Monitoring record sheet of mine surface deformation.

(2) Ledgers provided by underground mining with rock movement and deformation monitoring records can be regarded as surface deformation monitoring.

(3) Warning signs are set up for geological disasters′ (safe production) hidden troubles.

31 Dynamically monitor wastewater and tailings

Standard score: Five

Rating illustration:

Five points are rewarded for dynamically monitoring mineral processing wastewater, mine water, tailings (gangue hills), dumps, waste rock dumps, dust, and noise (sound)

Evaluation methods: Check data and the site

Basis or standard: Dynamic monitoring record.

[Interpretation] This article requires dynamically monitoring tailings (gangue hills), dumps, waste rock dumps, dust, and noise.

1. Monitoring of Tailings (Gangue Hills), Dumps, and Waste Rock Dumps

Monitoring indicators for tailings (gangue hills), dumps, and waste rock dumps:

Solid waste types, annual emissions, cumulative stockpiles, sources, annual comprehensive utilized volume, solid waste piles′ hidden troubles, and land occupation.

The number and scale of tailings ponds, the annual number of tailings accepted, the harmful components, hidden troubles, and the annual comprehensive utilized volume of tailings.

Monitoring methods of tailings (gangue hills), dumping sites, and waste rock dumps:

Manual field investigation, measurement, and remote sensing technology.

Onsite evaluation or supporting materials:

(1) *Mine Geo-environmental Protection and Land Reclamation Plan* and review comments.
(2) Dynamic monitoring system for tailings (gangue hills), dumps, and waste rock dumps.
(3) Dynamic monitoring record chart of tailings (gangue hills), dumps, and waste rock dumps.

2. Dust and Noise Monitoring
Dust and noise monitoring indicators:
(1) Dust concentration, dispersity, and free silica content.
(2) Noise intensity, namely the acoustic pressure in the sound field; noise characteristics, namely the various frequency components of acoustic pressure.
Dust and noise monitoring methods:
Monitoring personnel regularly use dust and noise testing equipment.

Onsite evaluation or supporting materials:
Dust and noise detection record sheet and ledger.

3. Online Monitoring of Mine Water and Mineral Processing Wastewater
Wastewater online-monitoring system is used for continuous real-time monitoring and remote supervising the purified sewage water quality. The integrated monitoring system, based on automated monitoring technology with online automatic analysis instrumentation as the core, can realize continuous real-time monitoring and remote supervising of water quality. Meanwhile, the system is conducive to the supervision of environmental protection authorities: grasp the water quality status of coal mine pollution sources in the jurisdiction, warn and forecast water pollution accidents, treat water pollution accident disputes across administrative regions, and supervise the total amount control system's implementation.
Monitoring parameters include COD, ammonia nitrogen, total phosphorus, and permanganate. Meanwhile, the reserved interface can be used to extend monitoring according to user requirements, such as total phosphorus, total nitrogen, heavy metals, phenol, fluorine, and oil content.
Onsite evaluation or supporting materials:
Water quality monitoring records and ledger.

32 Reclamation areas' dynamic monitoring
Standard score: Five
Rating illustration:

Five points are rewarded for dynamically monitoring land damage, stable state, soil, and reclamation quality in the reclamation areas
Evaluation methods: Check data and the site
Basis or standard: Dynamic monitoring record

[Interpretation] This article is a requirement for dynamically monitoring reclamation areas.

1. Land Reclamation Monitoring Should Meet the Following Requirements

    (1) Monitoring should be systematic and comprehensive. Land reclamation involves many disciplines. Therefore, the monitoring content should include the implementation scope and quality progress of the reclamation project, land damage, and eco-environment′ restoration, which ensures the available land in this area.
    (2) Monitoring plan should be classified and feasible. China's regional natural environment presents zonal characteristics, and land reclamation engineering measures are analogous. Therefore, land reclamation monitoring plans should be formulated according to the natural environment and construction projects' characteristics.
    (3) Monitoring settings should be optimized. The arrangement or setting of reclamation monitoring points, contents, and frequency shall adopt scientific and technical methods and reduce construction units′ unnecessary expenditures through reasonable optimization.
    (4) Monitoring standards should be based on the relevant national technical standards: *Technical Standards for Land Reclamation* (Trial), *Technical Specifications for Soil Environmental Monitoring* (HJ/T 166—2004), and *Technical Specifications for Wastewater Monitoring* (HJ/T 91.1—2019).

2. Monitoring of the Original Landform Surface Condition in Reclamation Areas

    (1) Initial landform information. Open-pit mines and underground mines will change the topography: Open-pit mining will form large pits and dumps, while underground mining will cause surface subsidence. Both mining methods cause topography changes, and mining is continuously changing. Therefore, it is necessary to detect the initial landform before mining for comparison.
    (2) Land use status. The original land use status information is retained, and the later changes are tracked and compared. It is based on land use data.
    (3) Soil information, including soil type and physical and chemical properties of soil information.
    (4) Settlement information. Collect information about villages and residents that need to be moved before mining.

(5) Cultivated land ownership. Collect cultivated land occupation and cadastral information in reclamation areas to provide a basis for the requisition-compensation balance.

3. Land Damage Prediction
Monitor land damage, such as excavation, collapse, and occupation. A land reclamation plan is taken as an example to illustrate the monitoring design of the subsidence area.

(1) Monitoring method. Measuring surface movement with leveling. Using the Yellow Sea Elevation System in 1980 to check and measure instruments and rulers before the operation. Besides, the measurement adopts the central wire method to read the data directly. The observation adopts the back-back-front order, with the accuracy reaching the third-class and the error in the observation less than 25 mm/km.
(2) Layout and establishment of the benchmark. Benchmark is the starting datum point for surface deformation monitoring. It is designed to set up two benchmarks on the road outside the mining area, using the second-class leveling data to measure the elevation and regularly checking the control points' stability.
(3) The layout of surface deformation datum marks. Eight sidings are laid along the coal seam′s strike and dip, with a total length of 80 km, and ten observation spots are set in the subsidence area. Deformation observation points and datum marks constitute a deposition monitoring network measured according to the fourth-level benchmark requirements.
(4) Monitoring personnel and frequency. Enterprises should entrust qualified professionals to monitor in time. The monitoring frequencies of benchmark and surface deformation are once every two months, while surface deformation monitoring points are once a month. Observation records should be accurate and reliable and organized timely to be compared with predicted results.
(5) Monitoring period. The reclamation plan′s length of service determines the specific monitoring period. The reclamation plan′s length of service is 42 years (2011–2052), including the construction period (3 years), the mining period (27 years), the stable settlement period (6 years), and the rehabilitation plant management and protection period (6 years). If it is determined to monitor the middle two periods, the monitoring period for surface subsidence is 33 years (2013–2045).

4. Monitoring of Reclamation Effect

(1) Soil quality monitoring
The natural characteristics′ testing content of land reclaimed for agriculture, forestry, and animal husbandry include reclamation areas′ topographic slope, effective soil thickness, soil available water, soil bulk density, pH, organic matter content, available phosphorus content, total

nitrogen content, and soil erosion modulus. The monitoring method is subject to the *Technical Standards for Land Reclamation* (Trial), with the monitoring frequency of at least once a year.

(2) Monitoring of reclamation vegetation
Reclaimed forest land monitoring includes plant growth potential, height, planting density, survival rate, canopy density, and growth. Meanwhile, vegetation monitoring content of reclaimed grassland includes plant growth potential, height, coverage, and grass yield. The monitoring method adopts a quadrat random survey, which shall be monitored at least once a year within the reclamation plan's service life, and at least once every three years after the reclamation project's completion.

(3) Monitoring of supporting facilities for reclamation
Auxiliary facilities for land reclamation include water conservancy engineering facilities and transportation facilities. The former includes irrigation, drainage, and related power facilities, while the latter includes roads at all levels and new footpaths between fields. Supporting facilities′ monitoring is subject to the land reclamation plan's design standards, including the completeness of newly-built supporting facilities, the guarantee of effective utilization, the restoration of damaged auxiliary facilities, and the satisfaction of the production and living needs of residents. Supporting facilities are monitored at least once a year.

5. Dynamically Monitoring Reclamation by Remote Sensing Technology
The multi-platform, multi-temporal, and multi-resolution characteristics of remote sensing data can expand mine-land reclamation monitoring from mines to mining areas and even the whole country. Therefore, the establishment of remote sensing dynamic monitoring system with the three-level from mines to mining areas and even national can monitor land reclamation and ecological reconstruction in mines across the country. The monitoring process includes remote sensing data collection and processing, data storage and management, data analysis and application, and achievement expression and display.

Onsite evaluation or supporting materials:

(1) Monitoring plan or system.
(2) *Mine Geo-Environment Management and Land Reclamation Plan* and acceptance opinions on mine geo-environment management.

Chapter 6
Comprehensive Utilization of Resources

The comprehensive utilization of resources has promoted the circular economy in the mining industry, including the development of associated resources, the recovery of valuable elements, and the resource utilization of waste. The principle of "resource, reduction, and harmlessness" realized new economic growth for enterprises and significantly reduced environmental pollution. It is an essential manifestation of the coordination of resource development and ecological protection.

The assessment focus of this chapter should reflect the comprehensive utilization of resources.

1 Comprehensive Utilization of Associated Resources

This section applies to non-metal, chemical, gold, metallurgy, non-ferrous metals, petroleum, coal, and other industries.

33 Resource exploration, evaluation, and development

Standard Score: Ten

Grading description: Ten points are rewarded for comprehensive exploration, comprehensive evaluation, and comprehensive development of associated resources according to the mineral resource development and utilization plan

Evaluation method: Search information

Basis or standard: Mineral resources development and utilization plan, and relevant product information

J. Deng et al., *Interpretation of Green Mine Evaluation Index*,
https://doi.org/10.1007/978-981-16-5433-6_6

[Interpretation] This article is a principled requirement for developing associated resources to improve the development and utilization of associated resources.

Mineral resources in many provinces in China account for a large proportion of economic development. With the issuance of the *Guiding Opinions on Establishing a Natural Reserve System with National Parks as the Main Body*, the development of a large number of mineral resources is limited, and it is imperative to develop low-grade associated resources.

Paragenetic resources, also called paragenetic minerals, refer to a group of minerals formed in the same origin and at the same metallogenic stage. Associated resources refer to minerals that are not in the same group but coexist in the same space with the same origin and the same mineralization period (or mineralization stage).

Paragenetic minerals are two or more minerals that meet industrial indicators and have at least a small scale (including small scales) in the same mining area (or deposit). Associated minerals are ores, minerals, or elements produced with the main mine and paragenetic ore in the ore deposit (or ore body). They do not have independent mining value but can be reasonably mined, extracted, and utilized simultaneously during mining and processing the main minerals.

1. Characteristics of Associated Resources
   (1) Most of the main minerals which have large quantities own associated resources.
   (2) Associated minerals are generally produced with the main minerals and have no technical and economic value for independent mining.
   (3) To ensure the development of the main minerals during the mining process, the impoverishment of associated resources is relatively severe.

   Based on the above characteristics, it is necessary to increase technological innovation and choose advanced, applicable, economic, and reasonable processes and technologies to utilize associated resources comprehensively. The extraction of low-grade valuable elements is the key, focus, and crux in developing and utilizing associated resources.

2. Comprehensive Exploration, Evaluation, and Development
   The comprehensive utilization of mineral resources is an extension of comprehensive exploration, evaluation, and development. Comprehensive prospecting, exploration, evaluation, and development are the basis for developing and utilizing associated resources. Comprehensive evaluation and exploration should be adopted for associated resources. Moreover, the following methods are used to develop associated resources, including simultaneous mining and transportation with main minerals, separate mining and transportation, and joint mining with main minerals and separation.

(1) Comprehensive exploration.
Comprehensive exploration is the simultaneous exploration of associated minerals and paragenetic minerals while exploring major minerals. It can improve resource utilization and economic benefits and reduce exploration costs. These are all issues that should be paid attention to in mineral exploration work.
1. During the comprehensive exploration, consideration should be given to associated components and their distribution rules, reasonable selection of sampling and testing methods, correct determination of industrial indicators, and selection of calculation methods of associated minerals' component reserves. Comprehensive exploration can improve resource utilization and economic benefits and reduce exploration costs.
2. The paragenetic minerals should be comprehensively explored and evaluated together with the main minerals. When some paragenetic minerals meet industrial utilization requirements, and their reserves reach medium-sized or above, the main minerals' exploration level cannot satisfy the corresponding conditions for paragenetic minerals. According to the minerals' actual needs and prospecting specifications, the exploration or unique prospecting work can be appropriately added.
3. The exploration of allogeneic symbiosis minerals should utilize the project of exposing the main minerals or increase the appropriate workload to investigate and evaluate the ore body. For situations where the scale of mineral resources reserves reaches medium-sized and above, exposing the main minerals cannot reach the corresponding degree of control. It is necessary to carry out unique exploration and evaluation work. Similarly, these tasks should only be added based on the actual needs and the mineral's exploration specifications.

(2) Comprehensive evaluation.
A comprehensive evaluation is a scientific and thorough evaluation of the main minerals, paragenetic or associated minerals found in an exhaustive exploration to determine their available economic and technical value and mining conditions.

During the comprehensive evaluation, the mineral composition, grain size, structure, and structural characteristics of the associated minerals should be verified for the various ores in the deposit. Besides, the occurrence status of useful, beneficial, and harmful components and the paragenetic relationship between minerals should be checked. For the components that exist in a dispersed state (such as in the form of isomorphism, inclusions in a solid solution or microcrystalline dispersion state, adsorbed on the surface of the mineral or colloid in the state of ions or complexions), the carrier mineral and its occurrence form should be ascertained. It is necessary to strengthen technological mineralogy research and choose reasonable processing and smelting methods and procedures. A comprehensive evaluation of the identified associated resources should be made in the exploration report.

(3) Comprehensive development.
Comprehensive development refers to the overall planning of paragenetic and associated minerals, and in a certain order, the simultaneous exploitation of different mineral deposits or various beneficial components of the same deposit, as well as the paragenetic and associated minerals at different levels.

(4) Relevant requirements.
   (1) Comprehensive exploration should be carried out when associated resources are found during the exploration process.
   (2) A comprehensive evaluation of the identified associated resources should be made in the exploration report.
   (3) In the resource development and utilization plan, comprehensively develop valuable associated resources.
   (4) What is needed is to determine whether the utilization of associated resources has development and utilization value under the current social, technical, and economic indicators and to make rational use in actual exploration and production.

3. Development and Utilization Plan of Associated Resources
Developing and utilizing mineral resources is to strengthen the early management of mineral resources' development and utilization. The technical plan that applicants need to submit to the mining registration management agency for mining rights is important for the mining rights review. The plan for developing and utilizing mineral resources should be compiled under the Ministry of Natural Resources' content. It includes an overview, mineral product demand analysis, mineral resources overview, main construction plans, mineral deposit mining, mineral processing and tailing facilities, environmental protection, land reclamation, and the brief conclusions of development plans and drawings.

When a plan is formulated for the comprehensive development and utilization of associated resources, it should be prepared and strictly implemented following of the *National Comprehensive Exploration and Evaluation Standards for Mineral Resources* and the *Contents of the Preparation of Mineral Resources Development and Utilization Programs*. The specific content includes an overview, mineral product demand analysis, mineral resource overview, main construction plan determination, mineral deposit mining, facilities of mineral processing and tailings, environmental protection, land reclamation, and the brief conclusion of development plan and drawings. For determining the comprehensive development and utilization plan of associated mines, considerations should be given:

(1) The mining design plan needs to consider the co-production of associated resources and use the existing production system.

(2) For projects with comprehensive utilization conditions, the project proposal, feasibility study report, and preliminary design should include associated resource development and comprehensive utilization of resources.

Onsite evaluation or supporting materials:

Mineral resources development and utilization plan, prospecting report, evaluation report, or development report or plan.

34 Comprehensive utilization of associated resources

Standard score: 20

Grading description: Use advanced, applicable, economic, and reasonable process technology to process and comprehensively utilize the associated resources; 20 points are rewarded for meeting the requirements

Evaluation method: Search information, and check the site

Basis or standard: Production report or financial report

[Interpretation] This article is a requirement for the mining technology of associated resources.

(1) Most of the main minerals with large quantities have associated resources.
(2) Associated minerals are generally produced together with the main minerals without technical and economic values for independent mining.
(3) To ensure the development of the main minerals during the mining process, the impoverishment of associated resources is relatively severe.

Based on the above characteristics, it is necessary to increase technological innovation and choose advanced, applicable, economic, and reasonable processes and technologies to comprehensively utilize associated resources. The extraction of low-grade valuable elements is the key, focus, and crux in developing and utilizing associated resources.

Onsite evaluation or supporting materials:

(1) Proof of advanced and applicable technology.
(2) Relevant approaches and facilities.

35 Comprehensive utilization of complex, refractory, or low-grade ores

Standard score: Five

Grading description: For complex and refractory to process or low-grade ores, it is necessary to adopt new processes to reduce energy consumption or a combined approach of beneficiation and metallurgy to improve technical and economic indicators. Five are rewarded points for achieving results and providing evidentiary materials

Evaluation method: Search information, and check the site.

[Interpretation] This article requires the comprehensive utilization of complex, refractory, or low-grade ores.

1. Complex and Refractory Beneficiation
   There are widely distributed and different mineral resources and secondary resources in nature. The definition of complex and refractory ore properties is closely related to the ore's properties and associated with the technological level of people's processing and utilization of mineral resources in different periods. It is also related to economic and environmental restrictions. Therefore, this definition is a relative concept. The same kind of ores, in different deposits or different mining areas, have various processing difficulties. The same type of minerals may have a large difference in selectivity in different countries and regions. Even if the ores in the same mining area at different times, the challenge of using ore will also change due to different technical levels.

   Complex ores generally have the following essential characteristics:

   (1) From ore properties, the complex ore has the characteristics of "poorness, fineness, and miscellany," such as the low grade of valuable elements, the fine grain size of useful minerals, and complex ore composition. It also has the characteristics of high mutual content of concentrates, high concentration of impurities in concentrates, low comprehensive utilization of associated resources, and low utilization of backwater.
   (2) From minerals' perspective, there are minerals with complex and refractory ores' properties, e.g., low-quality coal and refractory coal slime in coal resources; siderite, limonite, extremely fine-grained iron ore, oolitic hematite, and other low-quality iron ore in iron ore resources; mines with a complicated paragenetic relationship, high oxidation rate, high binding rate, and high-leaching-acid consumption in copper resources; low-grade, many calcium-containing impurity minerals, and tungsten-molybdenum resources that are similar in quality and accompanied by pyrrhotite in tungsten resources; low-aluminum–silicon ratio bauxite resources; low-grade high-silicon high-calcium magnesite, refractory rare and precious metal resources, rare earth mineral resources and tailings in magnesium and rare mineral resources.
   (3) From economy and environmental protection, due to the high cost, low benefit, and high pollution during the minerals' processing, this type of ores do not meet the economic and environmental requirements. Hence, such ores also belong to the category of complex ores.

With the development of technology, resource utilization improves, and the technical bottleneck of complex ore is overcome continuously. As a result, the complex ores' attributes accordingly change.

2. Low-Grade Ore
Low-grade ores refer to mineral resources that cannot be exploited alone due to the low grade of the main component under current economic and technical conditions. According to the grades, resources below the economic grade and above the boundary grade are collectively called low-grade mines. Moreover, the resources below the marginal grade and above the boundary grade belong to absolute low-grade mines, and the resources below the economic grade and above the marginal grade belong to relatively low-grade mines.

    (1) Cut-off grade.
    The cut-off grade refers to the lowest grade of that part of the resources that cannot be used under the conditions of a sharp increase in the price of mineral products, a sharp drop in the cost of mining enterprises, or strong government support, instead of under certain economic and technological conditions. The cut-off grade is the boundary between minerals and non-minerals. Its minimum requirement is greater than the grade of tailings.
    (2) Marginal grade.
    Marginal grade refers to the grade at which the profit of the produced mineral product is equal to the variable cost under the conditions of the mineral product price and production cost in orebody delineation (according to industrial indicators' requirements, the work of encircling the reserve calculation boundary on the reserve calculation drawing is called ore body delineation). Under this grade condition, mining enterprises can only maintain production without profits.
    (3) Economic grade.
    Economic grade refers to the minimum grade (break-even point) required for mining enterprises to recover their investment under orebody delineation's economic and technical conditions. That is, the production in this grade can just recover the investment. Additionally, the grade is the boundary between economic reserves and marginal economic reserves. Economic reserves are also the reserves delineated by the grade.

3. Innovative Technology

    (1) Using direct reduction and other combined beneficiation and metallurgy process.
    "Direct reduction" specifically refers to the process in which solid carbon below the blast furnace reflow zone and trivalent iron or divalent iron in the melt directly undergo a reduction reaction to produce metallic iron. "Smelting reduction" refers to the process of reducing iron ore in a high-temperature molten state without using a blast furnace to obtain liquid molten iron similar to blast furnace molten iron. Direct reduction and

smelting reduction have higher requirements for raw materials: usually high-grade lump ore, iron concentrate, or oxidized pellets.

The separation and reduction technology of complex refractory iron ore and slag iron put the complex refractory iron ore that cannot be directly used as the blast furnace's raw material in a higher temperature and stronger reducing atmosphere (some need to use reaction aids or other additives) than the magnetic roasting. Then reduce the trivalent iron or divalent iron in the ore to elemental iron. Simultaneously, the elemental iron should be grown to a certain size of metallic iron particles (metallic iron). This technology can facilitate the subsequent metallurgy process.

(2) Adopt new technology to reduce energy consumption and improve technical and economic indicators.

4. Complex and Refractory Ores

There are the following types of complex and refractory ores:

(1) Lead–zinc oxide ore has a wide variety of minerals and a complex ore structure.
(2) Gangue is complex and refractory iron ore such as fine-grained red (magnetite) iron ore with iron-containing silicate embedded, siderite, limonite, and oolitic ore.
(3) Low-grade refractory fluorite ore.
(4) Efficient separation of complex and refractory tungsten ore.

Onsite evaluation or supporting materials:

(1) Mining-process technology design or independent technological innovation of complex and refractory ore.
(2) Relevant approaches and facilities.

36 Requirements for associated minerals that cannot be exploited temporarily

Standard score: Five

Grading description: Five points are rewarded for effective protection measures for the associated minerals that cannot be temporarily exploited

Evaluation method: Search information

Basis or standard: Plan for development and utilization of mineral resources

[Interpretation] This article requires effective protection measures for the associated minerals that cannot be exploited temporarily.

Article 30 of the *Mineral Resources Law of the People's Republic of China* stipulates: During mining the main minerals, the paragenetic and associated minerals with industrial value should be planned in a unified manner, mined comprehensively, and utilized thoroughly without wasting. Effective protection measures should be taken to prevent loss and damages to minerals that temporarily cannot be mined comprehensively or that must be mined with main minerals but cannot be comprehensively utilized temporarily. Similarly, tailings containing useful components should be covered by such protection.

The minerals that cannot be comprehensively mined temporarily mean the associated minerals that have not yet been solved in mining, beneficiation, and smelting technology; the mined minerals that cannot be processed technically; minerals that are costly to be produced or temporarily not demanded in the domestic market; the associated minerals that can only be mined and transported together during the main minerals' mining; the main minerals and associated useful components that exist in the same ore and can only be mined at the same time, and separated in the process of separation and smelting.

Tailings refer to the product's part with the lowest content of beneficial ingredients in the beneficiation products. Among them, the valuable components are reduced to the extent that they cannot be recovered under the technical and economic conditions. This tailings is called the one containing useful components.

Mining enterprises should implement the policy of "Comprehensive Exploration, Comprehensive Evaluation, Comprehensive Mining, and Comprehensive Utilization" and strengthen feasibility studies on mineral resources' comprehensive utilization. New or rebuilt mines should also implement the technical plans, processes, and measures for comprehensive development and utilization. While mining and processing major minerals, mining enterprises and their superior authorities must make a unified plan: associated mineral resources should be comprehensively utilized with an industrial value under technically feasible and economically reasonable conditions. For minerals under inter-departmental jurisdiction, the division between departments and industries should be broken. It is needed to operate multiple industries surrounding the only main business: the department that mines the main minerals should take responsibility, and the relevant departments should participate in the planning and design. For important and complex mineral deposits, the state ought to specifically organize scientific and technological research on minerals' comprehensive utilization.

Adequate protection measures should be taken to minerals that temporarily cannot be mined comprehensively or that should be mined with main minerals but cannot be comprehensively utilized temporarily. Furthermore, tailings containing useful components should be covered by such protection. For minerals that temporarily cannot be exploited comprehensively, necessary security pillars can be retained, and roadways and essential facilities can be maintained for future mining. It is the mining enterprises' responsibility to implement the policy of combining comprehensive utilization with pollution control. These companies must also actively study their utilization and strengthen the management of unsold ore, fine ore, medium ore, tailings, waste rock, and coal gangue.

Article 18 Mining Technology introduces filling mining technology to protect minerals that cannot be thoroughly mined temporarily.

Onsite evaluation or supporting materials:

Mineral resources development and utilization plans or other systems can reflect the protection of associated minerals being temporarily incapable of exploitation.

2 Solid Waste Disposal and Comprehensive Utilization

This section applies to non-metal, chemic al, gold, metallurgy, non-ferrous metals, petroleum, coal, and other industries.

37 Industrial solid waste disposal and utilization
Standard score: 25

Grading description: Twenty-five points are rewarded for establishing comprehensive utilization of solid waste such as waste rock (slag), coal gangue, tailings, drilling waste mud, rock cuttings, scum, and oil sludge, and making full use of solid waste through backfilling, paving, and production of building materials

Evaluation method: Search information, and check the site

Basis or standard: Mineral resources development and utilization plan, and other evidentiary materials

[Interpretation] This article requires the comprehensive utilization of solid waste.

The comprehensive utilization of solid wastes such as waste rock (slag), coal gangue, tailings, waste drilling mud, rock cuttings, scum, oil sludge, and other solid wastes reduces environmental pollution and ecological damage during mining.

1. Downhole Backfill
   In the mining process, solid mine waste is used as the filling material in the goaf to realize the recycling and utilization of resources and reduce the damage to the ecological environment. There are two main methods: to backfill the mined-out area directly and crush the waste rock to a specific size. The former directly backfills the solid waste into the mined-out area without the waste rock produced by mining or the silt in the sump. The latter uses the crushed rock as a filling aggregate to improve all tailings' gradation and fill it into the goaf downhole.
2. Road Construction
   The waste rock and tailings contain a large amount of gangue, hard in texture, and a natural sandstone material. They can be used to make qualified sand and gravel aggregates and used as roadbed paving materials. On the one hand, it can significantly reduce the environmental impact of sand and gravel produced by

blasting the mountain; on the other hand, it solves the environmental problems caused by waste rock and tailings' surface storage.

Besides, for low-grade iron ore, high-efficiency dry and wet magnetic separators can also be used for pre-selection and polishing. It is required to make a small amount of pre-selected concentrate enter the rear mill for fine grinding and beneficiation. Many tailings thrown by dry and wet methods can be used as construction materials such as artificial sand and gravel, and non-burning bricks. This method allows pre-throwing tailings to be used and reduces the cost of beneficiation.

3. Produce Building Materials
   Mine solid waste can be used to produce building materials, and there are mainly the following four types:

   (1) Used to make cement and silicate building materials: Since solid waste contains many elements such as aluminum and silicon, silicate building materials can be made after extraction. The use of iron ore tailings and waste rock to produce concrete aggregate has advantages, significantly reducing the cement clinker and improving concrete performance. In the completed small-scale engineering tests, the amount of cement clinker in concrete can be reduced by 50 to 90%, and the durability of concrete can be significantly improved.
   (2) Glass, the ore contains much quartz, and the silicate ore containing a lot of fluorites, calcite, and dolomite. Granite-type tailings can be used as raw materials and ingredients for glass production.
   (3) Manufacturing of glass–ceramics. Glass–ceramics can build partition walls of houses, and its heat resistance and energy-saving type are relatively excellent.
   (4) Stone casting. The solid waste of mines used as an ideal material for cast stone contains granite, dolomite, and fluorite.

4. Utilization Requirements of Different Minerals

   (1) The construction stone mines in the construction stone centralized mining area should be equipped with advanced machine-made sand production lines. Multi-stage sedimentation and recycling are adopted to wash and drain solid waste such as stripped dirt that can be washed and sorted. The deposited dirt is used to make building bricks, basically achieving zero discharge.
   (2) Wastewater of metal-mine beneficiation should be fully recycled. The tailings of metal mines and coal gangue from coal mines should be utilized following the requirements of "Reduction, Resource Utilization, and Reuse."
   (3) Metal-mine tailings should be processed by processing-tailings fine powder as cement mixture, concrete, and mortar admixtures. Tailings can also be used to make aerated building bricks.

(4) Coal mines should use coal gangue to produce construction bricks and strive not to increase coal gangue emissions and gradually digest the stock of coal gangue. Finally, the comprehensive utilization of tailings and coal gangue is basically realized.

5. Principles of Comprehensive Utilization of Solid Waste
The primary way to control the environmental pollution of solid waste follows recycling, reduction, and harmlessness.

(1) Recycling of resources.
Recycling of solid waste to recover energy and resources should be utilized. The specific industry's production characteristics must determine the reclamation of industrial solid waste. Attention should also be paid to factors such as technical feasibility, product competitiveness, and economic benefits.

(2) Harmless disposal.
The harmless disposal of solid waste refers to proper treatment that makes the solid waste or its harmful components no longer dangerous to the environment or transforms the waste into substances that are harmless to the ground. Commonly used methods are incineration, composting, plasma gasification, and pyrolysis gasification.

(3) Reduction processing.
The amount of solid waste is significantly reduced through treatment.

Onsite evaluation or supporting materials:

(1) Resource development and utilization plan (comprehensive utilization process or system).
(2) Financial statements.

38 Topsoil disposal and utilization

Standard score: Ten

Grading description: Ten points are rewarded for stripping topsoil and coal overlying rocks for land reclamation and ecological restoration (no topsoil and overlying rocks will not be scored, and Article 37 Industrial solid waste disposal and utilization will be assigned thirty-five points)

Evaluation method: Search information, and check the site

Basis or standard: Mineral resources development and utilization plan and other evidentiary materials

[Interpretation] This article requires the disposal and utilization of topsoil.

Before mining, the cultivated land, forest land, and pasture to be destroyed should be stripped and stockpiled, and the stripped topsoil should be used for the reclamation of the damaged land.

Topsoil stripping refers to the stripping of the topsoil and rocks covering the upper part and surrounding of the ore body in open-pit mining, and then it needs to be transported to a particular place for disposal. During constructing green mines, the topsoil is required to be stripped and preserved for later land reclamation.

1. Topsoil Stockpiling

    (1) "Centralized Stockpile in Partitions and Fragments": the temporary stockpile of topsoil should occupy the free space on the site as much as possible. If there is no suitable pile on the site, additional land should be acquired, and temporary protective measures should be provided during the topsoil preservation. The nutrient-rich humus soil on the surface should be excavated and stacked in a concentrated manner.
    (2) The original ground should be leveled at the topsoil stacking point. In the process of topsoil backfilling and site preparation, the ground and surrounding topography should be coordinated. The low in the middle and high in the surroundings should be avoided to prevent soaking in depressions caused by rainy days.
    (3) The stacking point should be away from the river. If the topsoil is stacked in a slag yard, it should generally be stacked downstream of the slag yard or in a flat place on both sides, avoiding low-lying and water confluences.
    (4) For the topsoil dump, avoid high filling and carry out the layered filling and layered compaction.
    (5) If the amount of piled-up topsoil is small, it can be covered with plastic colored stripe cloth or film and pressed with soil bags surrounding it.
    (6) If the topsoil has a long storage period and exceeds one growing season, grass seeds can be sown for temporary greening. Moreover, the grass seeds should be selected forages with fertility (such as legumes).
    (7) After the temporary land occupation, the surface mud and stone layer should be removed first, and then the topsoil should be backfilled for comprehensive land preparation. After comprehensive land preparation, the ground height should be consistent with the surrounding area to restore greenery and re-cultivation (garden).
    (8) Soil covering is not required for spray seeding, soil nail hanging, net spraying, or grass greening.
    (9) Topsoil should be used for reclamation and greening or resource utilization.

    Note: There is a difference between roof cutting and topsoil stripping, though they both aim at open-pit mining. Mine roof cutting is a working platform made in advance for relatively large open-pit mines using top-down sub-station mining. The roof cutting part may be topsoil or ore (such as limestone mines). Topsoil stripping generally refers to the removal of useless overburden in smaller mines to extract the ore. Depressed surface mining generally requires topsoil stripping.

2. Topsoil Utilization.
   Stripped topsoil is used for reclamation protecting the surface mellow soil resources from loss and waste effectively. Simultaneously, it can reduce the extra capital input caused by soil transfer during land reclamation, with stripping topsoil used. As a result, good soil fertility and high crop yields can be achieved. Additional advantages are reducing the maturation cost and ripening time of the outside soils. Besides, soil erosion can be minimized.
   (1) Topsoil's utilization is one of the essential technologies and open-pit mining methods. Open-pit mining should adopt the integrated technology of stripping, dumping, land reclaiming, and reclamation, and methods of benching and independent storage of topsoil. These methods can provide feasible conditions for later ecosystem restoration. The most economical way of open-pit mining with topsoil separates and piles exposed waste rock and topsoil. After the mining is completed, the waste rock field and open mining pit should be covered with soil and regreening.
   (2) A suitable topsoil storage site should be built (waste rock yard) according to requirements. The topsoil storage site should be reasonably arranged and stored in an orderly manner.

Onsite evaluation or supporting materials:

Topsoil reservoir's stacking situation should be checked regarding the topsoil utilization plan in the mineral resources development and utilization plan,

39 Recovery and extraction of valuable elements or useful minerals
Standard score: Five
Scoring description: Five points are rewarded for achieving extracting valuable elements or useful minerals from solid wastes such as tailings, coal gangue, and waste rock
Evaluation method: Search information, and check the site
Basis or standard: Production report, sales report, and financial report

[Interpretation] This article requires the extraction of valuable elements from solid waste.

Re-selection of tailings (gangue) or recovery of valuable elements (contains beneficial components and utilization of useful minerals) is one of the primary measures for tailings" (gangue) overall comprehensive utilization, including old tailings' reuse and newly produced tailings re-selection. Tailings' re-selection can reduce tailings dam construction and maintenance costs and save grinding, mining, and transportation costs. Furthermore, it can save more investment in equipment's development, so it is getting more attention.

Mine tailings have the dual characteristics of secondary resources and environmental pollution. The use of beneficial components and useful minerals can reduce the discharge of final waste and reduce environmental pollution by recovering valuable elements.

To realize the extraction and recovery of valuable elements, a technically feasible, economical, and reasonable solution should be adopted. The following valuable-element recovery technologies are relatively developed:

(1) Extract various valuable elements from the chlorinated-volatilization dust collection solution of tin tailings (full extraction method).
(2) Extract rare earth from phosphogypsum.
(3) Extract valuable metals from cobalt earth ore.
(4) Comprehensive recovery of valuable metals from nickel smelting waste.
(5) Extract enriched rare-earth elements from coal gangue.
(6) Wet extraction of valuable metals from copper-nickel sulfide ore tailings.
(7) Extract valuable metals from zinc-sulfur separation tailings.
(8) Extract various elements from pulverized coal.
(9) Comprehensive recovery of valuable metals from lead-antimony smelting water quenching slag.
(10) Extract and comprehensively utilize valuable metals from red mud.
(11) Extract valuable metals from molybdenum tailings.
(12) Recover valuable elements from lead–zinc slag.

The key to the extraction of valuable elements is sifting technology's upgrading.

Onsite evaluation or supporting materials:

(1) Design for solid waste's comprehensive utilization.
(2) Financial statement.

3 Wastewater Treatment and Comprehensive Utilization

This section applies to non-metal, chemical, gold, metallurgy, non-ferrous metals, petroleum, coal, and other industries.

40 Mining wastewater's disposal and comprehensive utilization

Standard score: 15

Grading instructions: ①Seven points are rewarded for equipping mining wastewater treatment facilities such as mine water, drain water, drilling wastewater, well washing wastewater; ② eight points for using clean and resource technology to achieve effective wastewater treatment

Evaluation method: Search data, and check the site

Basis or standard: Production report (scheduling report) or other evidentiary materials

[Interpretation] This article requires the disposal and comprehensive utilization of mining wastewater.

After collecting and processing mine water, such as pit gushing water, underground gushing water, and flat tunnel gushing water, the collected water should be used for production water in the wells and undergrounds, industrial sites, waste rock yards, and roads-dust suppression. The remaining unusable part should be discharged based on the standard.

Mining wastewater's disposal and comprehensive utilization should adhere to the 10-character policies and priorities of "monitoring, collection, reuse, and external discharge," which can significantly reduce the amount of treatment and external drainage. It also helps enterprises reduce production costs.

1. Mining Drainage

   (1) Types of mining drainage.
   Underground mining: The caving method aims at karst drainage in every roadway; the filling method aims at karst drainage and filling tailings with water.
   Open-pit mining: Rain drainage and a small amount of karst water in the open pit.
   Waste rock field: Leaching drainage (primarily rainwater).

   (2) Characteristics of mining drainage.
      (1) Largely affected by the climate: the drainage volume is massive in the rainy season and small in the dry season.
      Underground drainage: Relatively stable, with a slight increase in the rainy season; pollutant concentration changes relatively small, but the underground sump volume is limited.
      The drainage volume of the open pit: Unstably, the drainage volume in the rainy season increases sharply, and the concentration of pollutants is low; the drainage volume in the dry season is small, and the concentration of pollutants is high. However, the storage capacity of the open pit can be flexibly increased or decreased.
      Leaching drainage volume of waste rock yard: Unstably, the drainage volume in the rainy season increases sharply, and the concentration of pollutants is low; the drainage volume in the dry season is small, and the concentration of pollutants is high; there is no storage capacity.
      (2) Failing to meet the emission standards of local governments or countries and industries. For example, unorganized or natural discharge will pollute the environment and endanger humans, animals, and organisms' survival.

2. Mining Drainage Treatment

   (1) General production drainage.
   General production drainage mainly removes suspended solids.

Process: Regulating tank, sedimentation tank, and discharge or utilization.

Regulating tank: The adjustment volume is determined according to the water volume or the water drainage.

Sedimentation tanks: Commonly used ones are horizontal sedimentation tanks, mechanically stirred sedimentation tanks, and vertical settlers. The assessment index is whether the suspended solids content exceeds the standard (70 mg/L for the primary standard and 300 mg/L for the secondary standard).

Sludge discharge disposal: The sludges discharged from the regulating tank and sedimentation tank have high water content. After further concentration by the thickener, the underflow of the thickener can be filtered by pressure. The dry slag obtained after the press filter should be piled up in a suitable place (non-hazardous slag) or put into the filling station's sand bin. The slag and tailings are filled into the well together so that the thickener's supernatant liquid can be directly discharged.

(2) Excessive heavy-metal-ion drainage.

Excess heavy-metal-ion drainage primarily removes heavy metals and suspended solids.

Heavy metal sewage is the industrial sewage with the most severe environmental pollution and the greatest harm to humans. Heavy metals generally refer to metals with relatively high density (density ≥ 5). Specifically, it refers to metals with atomic numbers above 24 in the periodic table.

Sewage containing these heavy metals is discharged into the environment and impacts humans through the soil, water, and air. Certain heavy metals and their compounds can accumulate in fish and other aquatic organisms and crop tissues. These enriched heavy metals cause more extensive and more severe harm to humans through drinking water and the food chain. Since heavy metals are difficult to degrade and destroy, people pay more attention to heavy metal pollution sources, and the standards for wastewater treatment and discharge are becoming stricter. So far, no matter at home or abroad, heavy metal sewage treatment is not perfect and thorough. It is far from preventing heavy metal sewage from polluting the environment. The mercury, arsenic, lead, and fluorine pollution in some water bodies in China and the cadmium pollution in some sewage irrigation areas are still quite severe.

Primary sources of heavy metal sewage are the machinery processing industry, mining industry, steel and non-ferrous heavy metal smelting, and partial chemical industry production. It mainly includes heavy non-ferrous metal mine pit drainage, waste rock leaching water, concentrator tailings drainage, flue-gas dust-removal drainage of heavy non-ferrous metal smelting plant, floor washing drainage of wet smelting workshop (leaching, clean liquid, electrolysis), and plant's rain drainage. These drainages contain various heavy metals. Their drainage characteristics are related to the manufacturing technique, products' control degree, and managers' level. Advanced production technology is particularly prominent: it shows unstable drainage and unstable discharge of heavy metal ions, which brings specific difficulties to the latter's sewage treatment and compliance.

Once this substandard sewage is discharged into the water body, the impact on human survival and sustainable development will be massive. In the 1950s, Japan's shocking Minamata disease and Toinai disease were caused by mercury-containing sewage and cadmium-containing sewage, respectively. In 1973, China formulated the *Comprehensive Wastewater Discharge Standard* setting limits on the maximum allowable discharge concentration of the first and second pollutants. In 1988 and 1996, revision work was carried out to increase the first and second pollutant discharge indicators' limits.

Heavy metal wastewater treatment can be divided into two categories:

The first one converts heavy metals in a dissolved state into insoluble heavy metal compounds removed from the sewage by precipitation and floatation. Specific methods include neutralization, sulfurization, oxidation–reduction method, and ferrite. This method is widely used in China due to its low processing costs and vast medicine sources. It also adapts to water volume changes and has relatively simple operation and management.

The second one concentrates and separates heavy metals in sewage without changing their chemical forms. Specific methods include reverse osmosis, electrodialysis, evaporation and concentration, and ion exchange. Such methods are rarely used in China due to high processing costs and the difficulty of adapting to water volume changes.

Table 1 shows various heavy metal treatment technical solution's comparisons.

3. Mine Water

Mine water is primarily the mining equipment's cooling water, wet rock drilling water, and dust suppression water. These waters do not require high water quality, and the primary indicator is suspended solids. A secondary water system should be adopted to increase the wastewater reuse rate to use water resources and reduce new water. Mine gushing water is concentrated in the sump of the underground collection system. After sedimentation in the sedimentation tank, the suspended solids' content in the wastewater is reduced to 100 mg/L, which meets the mining process's water index requirements. This water is used for mining process water such as rock drilling water and water spraying to reduce dust.

The toilets in the mining area are dry latrines, and the feces are cleaned and used as fertilizer. The domestic sewage in the mining department is mainly produced by washing hands. After being collected and settled, it can be used to

Table 1 Comparison of technical solutions of heavy-metal treatment

Serial number	Program name	Brief description of the principle	The primary advantage	Major flaws	Investment situation	Comprehensive utilization, energy-saving, and environmental protection characteristics
1	Vulcanization method	Insoluble metal sulfide precipitates are generated to separate the metal ions according to the reactions of metal ions and sulfide ions	1. Easy to dehydration 2. More than 90% of heavy metals can be recycled 3. Less slag	1. Long process flow and high processing cost 2. It is easy to cause secondary pollution during operation 3. The content of sulfide ions in the drainage exceeds the standard and needs to be further neutralized by the lime method, but it is not easy to settle 4. It cannot form a single treatment system	High investment	Heavy metals are primarily recycled

(continued)

Table 1 (continued)

Serial number	Program name	Brief description of the principle	The primary advantage	Major flaws	Investment situation	Comprehensive utilization, energy-saving, and environmental protection characteristics
2	Lime method or iron salt lime method	For forming insoluble metal hydroxide precipitation, an alkaline neutralizer should be added to the heavy metal sewage to make the metal ions react with the hydroxyl groups according to the optimal pH for forming hydroxides of various heavy metals. Then the metal ions should be separated from the water. The addition of iron salt is mainly for the reduction and aggregation of heavy metals	1. The process is simple, and the treatment effect is good 2. Convenient operation and management 3. Low processing cost	1. Large amount of slag and high moisture content 2. The metal content in the slag is low and not suitable for recycling 3. The accumulation of slag can easily cause secondary pollution 4. It is easy to cause secondary pollution during the preparation of lime milk 5. The treated water contains high calcium and magnesium ions	Low investment	The slag contains heavy metals. Without keeping correctly, it is easy to cause secondary pollution

(continued)

Table 1 (continued)

Serial number	Program name	Brief description of the principle	The primary advantage	Major flaws	Investment situation	Comprehensive utilization, energy-saving, and environmental protection characteristics
3	Ferrite Method	The method adds iron salt and alkali to the sewage containing heavy metal ions and uses the co-precipitation method to prepare ferrite powder from the sewage	1. Low investment 2. Suitable for various heavy metal ions 3. The sediment is easy to dehydrate, and its chemical properties are stable	1. The processing cost is too high 2. It is hard to control 3. Domestic actual use is small, and it lacks test data	Low investment	No secondary pollution
4	Special industrial membrane separation technology	It is only used for wastewater recycling	1. Perform at room temperature 2. No phase change 3. No chemical change 4. Good selectivity 5. Strong adaptability	1. Products can be concentrated into dry matter 2. Isomers cannot be separated	Low investment	Low energy consumption

(continued)

Table 1 (continued)

Serial number	Program name	Brief description of the principle	The primary advantage	Major flaws	Investment situation	Comprehensive utilization, energy-saving, and environmental protection characteristics
5	Electrocoagulation-sewage treatment technology	Like the ferrite method, iron ions obtained by the iron plates' electrolysis collide with heavy metal ions to form solid particles. Then precipitate the particles out of water	1. High degree of automation and good processing effect 2. Convenient operation and management 3. No chemicals added 4. Lower processing cost	High influent pH and high pretreatment cost	High investment	Recyclable heavy metals are currently in the production experiment and promotion stage

green the mining area and the ground sprinkling to reduce dust and not enter the surface water.

4. Mine-Water Treatment Methods and Comprehensive Utilization
   The mining of mineral deposits destroyed the original occurrence of groundwater and created cracks. Due to the close hydraulic connection between atmospheric precipitation, surface water, groundwater, and domestic water, all the water penetrates the underground mining space along the original and new fissures to form mine water. Mine water is the most discharged wastewater during production. For a long time, mine water has been treated as a water hazard for prevention and treatment due to technical limitations and insufficient understanding. Therefore, it has been discharged in vain without comprehensive utilization and protection. With the development of science and the improvement of people's awareness of environmental protection, people have gained a new understanding of mine water: they treat mine water as a water resource, that is, mine water resource.

   (1) Mine-water utilization's necessity.
       Mining in the mining area drains a large amount of groundwater, destroys and drains the groundwater resources in the mining and surrounding areas, and causes the groundwater level to drop. These actions give rise to the depletion of water resources in the mining area, cause the ground drop in the hidden mining area, and induce the karst ground collapse in the karst mining area. If a large amount of mine groundwater is directly discharged, it will cause water quality deterioration and pollution. The groundwater is relatively clean when initially flowing into the shaft and roadway. If the mine's underground water resources are purified into drinking water, it can meet the production and domestic water and save many funds for drilling deepwater source wells. Besides, it can create better economic and environmental benefits. Most coal mines suffer from severe water shortages, so it is necessary to utilize mine water.
   (2) Mine water treatment technology.
       At present, following the degree of environmental impact and the feasibility of serving as a drinking water source, China has customarily sorted mine water into five categories according to the characteristics of water quality: clean mine water, mine water containing suspended solids, mine water with high salinity, acid Mine water and mine water containing toxic and harmful elements or radioactive elements. Different mine water adopts different treatment methods.
       A. Clean mine water.
          Clean mine water mostly refers to Ordovician lime water, sandstone fissure water, Quaternary alluvial water, and old goaf water. This kind of water is neutral in quality, low in turbidity and mineralization, and very low in toxic and harmful elements. It meets the standards of drinking water and can be used by special water pipelines. Furthermore, it needs to be disinfected when used as drinking water.

B. Mine water with suspended solids.
Suspended-solids mine water is generally black and contains many suspended solids such as coal particles, rock, and powder, but its total hardness and salinity are not high. The primary characteristic of suspended matter is that it is suspended in moving water but can be separated in still water, floating lightly and sinking heavy. According to suspended solids' characteristics, the primary methods commonly used for industrial water purification are coagulation and precipitation. Coagulation is an essential part of the water treatment process. The principle of choosing coagulant is producing large, heavy, and strong alum, with good water purification effect, no adverse impact on water quality, low price, and sufficient supply. Commonly used coagulants are aluminum salt and iron salt coagulants. The mixing process disperses the agent into the water quickly and evenly. It should be uniformly mixed with the raw water in the shortest possible time so that all colloidal impurities in the water can interact with the agent. The colloidal impurities in the water agglomerate into larger alum particles by adding the coagulant to the raw water and mixing. Then the solids should be removed in the sedimentation tank. Besides, drinking water standards can be reached after fast filtration and disinfection treatment.

C. Mine water with high salinity.
High-salinity mine water refers to mine water with a total mineralized inorganic salt content greater than 1000 mg/L. It mainly contains SO_i^{2-}, CI^-, Ca^{2+}, K^+, and Na^+ plasma. Most water quality is neutral or alkaline; the hardness is correspondingly high with a bitter taste; a few are acidic. High-salinity mine water is not conducive to crop growth and will salinize the soil. It is easy to scale when used as boiler water, and will affect the quality of concrete when used for construction water. Drinking it for a long time will cause diarrhea and indigestion, with more severe effects on patients with heart and kidney diseases. Mine water in water-deficient mining areas in northern China is often high-salinity mine water. Therefore, it is necessary to process into drinking water and production water through purification and desalination processes.

D. Acid mine water.
Acidic mine water refers to mine water with a pH of less than 5.5, generally ranging from 3 to 3.5, and individually less than 3, with high total acidity. When mining sulfur-bearing coal seams, sulfur is oxidized and biochemically reacted to produce sulfuric acid. Acidic water easily dissolves the metal elements in coal and its surrounding rocks, leading to increased iron, manganese, heavy metals, and inorganic salts and increasing the degree of salinity and hardness. Acidic water is ordinary in high-sulfur mining areas in southern China. Acidic water is easy to corrode mine equipment and drainage

pipelines and endanger the health of workers. If it is pumped to the ground, it will affect the soil pH, cause soil compaction and crop wilting, and increase the surface water acidity affecting aquatic organisms' survival directly. Acid mine water must be discharged after the treatment reaches the standard due to its severe damage to the environment and ecology.

E. Mine water containing toxic and harmful elements or radioactive elements.

Mine water containing toxic and harmful elements or radioactive elements refers to water containing fluorine, iron, manganese, copper, zinc, lead, uranium, radium, and other elements. Fluorine-containing mine water essentially comes from groundwater areas with high fluorine content or areas containing fluorine mineral fluorite (CaF_2) or fluorapatite in coal and surrounding rocks. Drinking high fluoride water is prone to osteoporosis, dental fluorosis, and other diseases. Some coal mine water in northern China contains more than 1 mg/L of fluorine. Mine water containing iron and manganese generally forms under groundwater reduction conditions, and most of it presents a low-priced state of divalent iron and manganese. It has a smell of iron and is easy to become turbid, reducing the dissolved oxygen of surface water. Therefore, this type of water needs to be treated before use or discharge. Mine water containing heavy metals principally refers to mine water containing copper, zinc, lead, and other elements. These elements' concentration meets the discharge standard but exceeds the standard for drinking water, so it is not suitable for direct drinking. Radioactive element water mainly refers to mine water containing natural radionuclides such as uranium and radium and their decay product radon that exceed the standard for drinking water. For this type of mine water, suspended solids should be removed first, and then the pollutants not up to the target water quality should be treated.

5. Mine-Water Disposal rate's Calculation

Disposal rate = the amount of mine water (drained water) treated and reached the standard / total amount of mine water (drained water) × 100%.

Regardless of reuse or discharge, it must be treated under the corresponding standards, and the disposal rate should reach 100%. After treatment, mine water and drain water can be used as recycling water for beneficiation water, water for dust removal, irrigation water for greening, and maintenance.

6. Mine Water Disposal Rate Requirements

(1) Coal industries require the disposal rate of mine water and drain water to reach 100%.

(2) Gold industries require the mine water disposal rate to reach 100%.

(3) Chemical industries require that the utilization (or disposal) rate of tailings water and brine in water-deficient areas in Northwest China is not less than 95%.

Onsite evaluation or supporting materials:

Mining-wastewater treatment design or mine-water treatment volume in production reports.

41 Production wastewater' 's disposal and comprehensive utilization
Standard score: 15
Grading description: ① Seven points are rewarded for establishing a recycling treatment system for production wastewater such as beneficiation wastewater; ② eight points for achieving production wastewater's recycling
Evaluation method: Search data, and check the site
Basis or standard: Production report (scheduling report) or other evidentiary materials

[Interpretation] This article requires the disposal and comprehensive utilization of mining wastewater.

1. Production wastewater's Source

(1) Crushing-mine dust-collection drainage and ground-washing drainage. This part of the drainage volume is not large, but many spots are dispersed in a vast collection area. Its central component is suspended solids, which are high in content and easily cause pipeline blockage. In the design, water pumps are generally used for pressurized transportation.

The processing method is as follows:

(1) A sedimentation tank should be set up onsite. The clear water after precipitation should be reused on the spot or sent to the backwater pumping station in a centralized manner, and the sediment should be manually cleaned by pumping to the grinding port of the ball mill. Its advantage is that the pipeline stroke is short and not easy to be blocked. However, it increases the workshop's labor-management cost.

(2) A thickener should be set up. Water from various places should be drained into the thickener for processing. The thickener's supernatant should be sent to the return pump station for centralized reuse, and the sediment should be pumped to the grinding port of the ball mill. Its advantage is convenient centralized water management. However, it needs to increase pipeline and equipment investment since each drainage point needs to be pumped to the thickener, and gravity flow is not suitable.

Adopting the above methods in line with local conditions can achieve good results.

(2) Concentrated thickener's drainage.

In non-ferrous beneficiation and black beneficiation, most concentrated thickener drainage can be directly reused without processing. However, in non-ferrous beneficiation, various beneficiation agents are added to increase a few metals' recovery rates. Some of them affect the beneficiation index after reaching a particular concentration. The surplus water will then be discharged from the beneficiation process and supplemented by new water produced. The discharged sewage needs to be treated and discharged following the standard.

(3) Tailings thickener's drainage.

Tailings thickener drainages (also called plant-front backwater) have a large amount of water and can be directly reused. Only when the cyanide method is used in gold beneficiation should the tailings slurry and water be treated together to meet the cyanide discharge standard before it can be discharged. The alkaline chlorination method is relatively economical.

(4) Tailing pond's drainage.

In the mine project, the tailings pond is the last outlet of the whole project. The quality of its drainage water will directly affect the water quality of local rivers, rivers, and lakes, closely related to the local ecosystem.

In the early beneficiation production, the tailings pond was the largest water storage tank for reuse. A large amount of water can be reused after the tailings slurry is transported to the tailings pond for natural clarification; in recent years, most of the tailings backwater has been changed to the plant-front thickener's backwater, significantly reducing the storage capacity of the tailings pond. Hence, the warehouse cannot store reused water anymore. If the pollutants' dissolution in the storage does not meet the discharge standard, it needs to be treated and discharged after reaching the standard.

(1) The pollutants at the outlet of the tailing ponds are related to the minerals selected for beneficiation: It primarily manifests in excessive pH, heavy metals, fluoride, and cyanide. Therefore, it sometimes fails to meet local or national and industry discharge standards.

(2) This part of the wastewater treatment process is the same as the previous treatment process, except for the treated water size.

(3) Tailings reservoir's clarified water and the tailings dam's seepage water should be recycled to the processing plant after being collected by the return pond. An accident water collection pond and a material storage reservoir should be provided under the dam, and standby power supply and return water facilities should be provided to ensure the tailings water will not be discharged during accidents.

2. Mineral-Processing wastewater's Treatment and Comprehensive Utilization
Beneficiation wastewater includes beneficiation-process draining water, tailing-pond overflow water, and mine draining water. The draining water from the beneficiation process is generally transported to the tailings pond and the tailing slurry, collectively referred to as tailing water. Therefore, the treatment of beneficiation wastewater is also called tailing-water treatment.

A large amount of discharged wastewater is one of the characteristics of wastewater from mineral-processing plants in China. Besides, beneficiation wastewater has a large water volume, high suspended solids content, many harmful substances, and a low concentration. The beneficiation water consumption per ton of ore is 5–10t. In 1973, China's mineral processing wastewater discharge amounted to one billion cubic meters. Moreover, the second characteristic of the wastewater from the concentrator in China includes relatively complex wastewater composition with toxic and harmful components and a low concentration.

(1) Mineral-processing wastewater's hazards.
Pollutants in mineral processing wastewater essentially include suspended solids, xanthate, black medicine, terpineol, cyanide, sulfide, chemical oxygen-consuming substances, and other contaminants such as oil, phenol, ammonium, and phosphine. The hazards of heavy metals such as copper, lead, zinc, chromium, mercury, and arsenic plasma and their compounds are well known and will not be discussed here.

The major hazards of other pollutants are as follows:

1) Suspended matter: Suspended matter in water can block fish gills and affect algae's photosynthesis, thereby disturbing aquatic organisms' living conditions. The river may be silted when suspended solids' concentration is too high, and irrigation will compact the soil. If it is used as domestic water, the suspended matter will make people feel uncomfortable. Additionally, it is also a carrier of bacteria and viruses, potentially harmful to the human body. More seriously, when heavy metal compounds are present in the suspension, they will be released into the water under certain conditions (the pH of the water drops, the ionic strength, and the concentration of organic chelating agents).
2) Xanthate: Xanthogenate is a light-yellow powder with a pungent odor and easy to decompose. Its odor threshold is 0.005 mg/L. The fish and shrimps in the water contaminated by xanthate have an unpleasant smell of xanthate. Besides, xanthate is easily soluble in water and unstable in water, incredibly easily decomposed under acidic conditions. Its decomposition product, carbon monosulfide (CS), can be a sulfur pollutant. Therefore, the maximum allowable concentration of butyl xanthogenate in surface water in China is 0.005 mg/L.

3) Black powder: Dihydroxydisulfide phosphate is the main component, and its impurities include formic acid, phosphoric acid, thiocresol, and hydrogen sulfide. It is a dark-brown oily liquid, slightly soluble in water, with the smell of hydrogen sulfide. It is also the source of phenol, phosphorus, and other pollution in mineral processing wastewater.
4) Terpineol oil: #2 flotation oil, the main component is terpene alcohol. Yellow–brown oily transparent liquid, insoluble in water, is a non-toxic mineral beneficiation agent but has a loose fragrance to cause changes in waterbody's perception. Terpineol oil is a foaming agent tending to cause unpleasant foam on the water surface.
5) Cyanide: A highly toxic substance. After entering the human body, it hydrolyzes into hydrocyanic acid with gastric acid and is absorbed by the intestines and stomach to enter the blood. The hydrocyanic acid in the blood can combine with the iron ions in cytochrome oxidase to produce oxidized-ferric-cytochrome-acidification enzymes. As a result, the blood loses its ability to transmit oxygen, causing tissue hypoxia and causing poisoning. However, cyanide can be removed by the self-purification of the water body. Therefore, it can meet the discharge standard if this feature is used to extend the beneficiation wastewater's retention time in the tailing pond.
6) Sulfide: Under normal circumstances, S and HS will affect the water body's sanitary condition in water and generate hydrogen sulfide under acidic conditions. When the hydrogen sulfide content in the water exceeds 0.5 mg/L, it is toxic to fish, and the omitted odor can be detected; hydrogen sulfide's odor threshold in the atmosphere is 10 mg/m^3. Besides, low-concentration CS is volatile in water and enters the human body through breathing and skin. Long-term contact can cause poisoning and lead to neurological disease: Charcot carbon-disulfide hysteria.
7) Chemical matters consuming oxygen: Chemical matter consuming oxygen is a quantitative and alternative indicator of oxygen-consuming organic matter in water. Oxygen-depleting substances in beneficiation wastewater are essentially beneficiation agents remaining in the water.

(2) Pollution-treatment method.

The following treatment units can be used for the mentioned pollution in wastewater.

Suspended matter: Pre-sedimentation and coagulation or sedimentation methods are mainly used.

Acid–base wastewater: Methods are wastewater's mutual neutralization and acidity' 's neutralization by tailing alkalinity.

Heavy metal ions: Methods are co-precipitation or flotation technology for adjusting the pH of raw water, sulfide precipitation, lime-flocculation precipitation, adsorption technology (including biological adsorption), chelating resin method, ion exchange method, and constructed wetland technology.

Xanthate and black powder: Methods are iron salt coagulation or precipitation technique, bleaching powder oxidation, Fenton oxidation degradation method, and constructed wetland technology.

Cyanide: Methods are natural purification technique, hypochlorite or liquid chlorine oxidation, hydrogen-peroxide oxidation method, iron-complex combination method, insoluble-salt precipitation method, acidification-volatile re-neutralization method, zinc sulfate-sulfuric acid method, sulfur-dioxide air-oxidation method, electrolytic oxidation method, ozone oxidation method, ion exchange method, biodegradation method, and constructed wetland.

Sulfide: Methods are mutual precipitation with heavy metal-containing wastewater, blowing-off method, air-oxidation method, chemical precipitation method, chemical oxidation method, and biochemical oxidation method.

Chemical oxygen demand: Methods are coagulation or sedimentation, biodegradation, advanced oxidation, and adsorption method.

(3) Comprehensive utilization.

A. Beneficiation wastewater's treatment by coagulation inclined-tube sedimentation method.

The workshop's wastewater should be separated from the solid and liquid through the grit tank, where the grit in is discharged into the tailings yard through the unloading gate. The supernatant overflowed from the grit tank enters the reactor to coagulate after mixing by dosing. Then it flows into the inclined-tube settler to remove the fine suspended solids and harmful substances. The sediment of the inclined-tube settler should be discharged to the tailings sand field through the valve. After this process, the wastewater reaches the national allowable discharge standard. According to environmental protection requirements, the effluent from the inclined-tube settler enters the clean water tank and is pumped back to the workshop for reuse with a clean water pump, saving water and closing the wastewater cycle to achieve zero discharge.

B. Coagulation sedimentation, activated carbon adsorption, and reuse process.

This method is the most widely used method for recycling wastewater from mineral processing plants in China. Experimental research on different mines' beneficiation wastewater found that the results of putting different reagents or the same amount of the same reagent into the same beneficiation wastewater are different. Nevertheless, the common points are as follows:

1) Coagulant-effect comparison test: Polyferric sulfate (PFS), mixed aluminum chloride (PAC), and alum are used as coagulation and precipitation agents. The results show that it is more

economical and reasonable to use alum as a coagulant, and its optimal dosage is at about 30 mg/L.

2) Polyacrylamide's (PAM) influence on the coagulation effect: PAM's addition further improves wastewater's coagulation treatment effect. However, it is an organic polymer increasing the COD value in water. In practice, considering the change of the coagulation treatment effect and the increase of the COD value, the PAM input amount can be set to 0.2 mg/L.
3) Settling time's impact on wastewater: The established standing time after coagulation is 30 min.
4) Adsorption test: The amount of powdered activated carbon is less than that of granular activated carbon, and the same effect can be achieved in half of the case. Simultaneously, it is selected as the adsorbent because the powdered activated carbon is easy to enter the concentrate and will not accumulate in the water cycle. The optimal dosage is generally 50–100 mg/L.
5) Flotation test: After coagulation precipitation and activated carbon adsorption, the wastewater can be reused without any influence on the mineral processing index. After coagulation and precipitation of alum (30 mg/L) and PAM (0.2 mg/L), and then purified by powder activated carbon (50–100 rag/L) process, the effluent quality meets the national mine wastewater discharge standards and brings benefits to enterprises considerable economic benefits. More specifically, the reuse results show that the wastewater treated by this process can be reused without affecting the beneficiation index and reduce the number of flotation reagents in the beneficiation process. Simultaneously, due to the reuse of wastewater, the daily amount of fresh water is reduced. It has the social significance of reducing pollution and purifying China's environment since China has water shortages. The method has a simple process, good effect, and a wide range of industrial application prospects.

C. Comprehensive method for mineral-processing wastewater's resource utilization.

After comprehensive research on water treatment experiments and beneficiation's comparative experiments, professionals have concluded a relatively feasible solution to mine beneficiation wastewater.

Due to the different quality of various wastewaters, the regulating tank plays a role in regulating water quality and volume in reuse and treatment. The coagulation-sedimentation tank can strengthen the mixing of the coagulant and the wastewater so that the fine particles can grow and become suspended solids that can be removed by sedimentation. The reaction tank is used to deepen wastewater treatment further, use defoaming agents to react to the excess foaming agent in the wastewater, and weaken the flotation index's influence.

Onsite evaluation or supporting materials:

Mining-wastewater treatment design or the amount of beneficiation-wastewater treatment in the production report.

42 Domestic sewage disposal

Standard score: Ten

Grading description: ① Four points are rewarded for equipping a domestic sewage treatment system; ② six points for effective disposal of domestic sewage

Evaluation method: Search data, and check the site

Basis or standard: Production report (scheduling report) or other evidentiary materials

[Interpretation] This article requires production wastewater's disposal and comprehensive utilization.

Domestic sewage in mine projects essentially comes from canteens, bathrooms, and office buildings. Because the drainage points are scattered, the domestic sewage treatment should be centralized as far as possible. If it is not suitable, the domestic sewage treatment facilities should be dispersed. The primary pollutants removed are indicators such as biochemical oxygen demand (BOD5), chemical oxygen demand (COD), suspended solids, and ammonia nitrogen. After treatment, the water quality needs to meet the local discharge standard, or *Integrated Wastewater Discharge Standard* (GB 8978–1996).

1. Approaches

   (1) Sewage from the living quarters and employees' catering should enter the collection tank with the washing wastewater after being treated by the grease trap. Finally, it should be used for dust suppression and sprinkling on industrial sites and roads.

   (2) If there are dry latrines, the fecal water should be cleaned and collected regularly for fertilizing the surrounding farmland. Suppose there are flush toilets in the office and living areas. In that case, the domestic sewage should be treated in septic tanks, cleaned and collected regularly, and then used to fertilize the surrounding farmland or be used in industrial sites, roads, dust depression, or discharged following the standard after being treated by integrated domestic-sewage treatment facilities.

2. Processing Method.
   Domestic sewage treatment methods: China has relatively mature domestic sewage treatment methods and national standard atlases. Besides, sewage-treatment equipment has developed manufacturers. The central techniques of domestic sewage treatment are as follows:

   (1) Physical treatment method.
   The physical treatment method refers to the wastewater-treatment process separating and recovering the insoluble pollutants in the suspended state (including oil film and oil droplets) in the wastewater through physical action. It can be classified into a gravity separation method, centrifugal separation method, and sieve filtration method. The treatment method based on heat exchange's principle is also a physical treatment method.
   (2) Chemical treatment method.
   The chemical treatment method refers to the wastewater treatment method separating and removing the dissolved and colloidal pollutants in the wastewater or converting them into harmless substances through chemical reactions and mass transfer. In the chemical treatment method, the treatment units based on the chemical reaction caused by the dosing of chemicals are coagulation, neutralization, and redox. The treatment units based on mass transfer include extraction, steam stripping, air stripping, adsorption, ion exchange, electrodialysis, and reverse osmosis. The last two processing units are collectively referred to as membrane separation technology. The processing unit that uses mass transfer has both chemical effects and related physical effects, so it can also be separated from the chemical treatment method and become another type of treatment method, called the physical–chemical treatment method.
   (3) Biological treatment method.
   The biological treatment method refers to the wastewater treatment method in which the organic pollutants in solution, colloid, and fine suspension state in the wastewater are transformed into stable and harmless substances by the metabolism of microorganisms. According to the different microorganisms, biological treatment can be divided into aerobic biological treatment and anaerobic biological treatment. The aerobic biological treatment method is widely used in wastewater biological treatment. Furthermore, the aerobic biological treatment method is divided into activated sludge and biofilm methods. The activated sludge method itself is a processing unit, and it has a variety of operating modes. The processing equipment belonging to the biological membrane method includes a biological filter, rotating biological disk, bio-contact oxidation, and biological fluidized bed. The bio-contact oxidation method is also called the natural biological treatment method. Anaerobic biological treatment, also known as biological reduction treatment, is essentially used to treat high-concentration organic wastewater and sludge. The processing equipment used is chiefly a digester.

(4) Bio-contact oxidation method.

The bio-contact oxidation method refers to the use of the bio-contact oxidation method to treat wastewater. Precisely, this method fills the biological reaction tank with fillers. Then the oxygenated sewage immerses all the fillers and flows through the fillers at a specific flow rate. The packing is covered with biofilm contacting with sewages. With microorganisms' metabolism on the biofilm, organic pollutants in sewage can be removed, and sewage can be purified. Finally, the treated wastewater is discharged into the bio-contact oxidation system and mixed with domestic sewage for treatment. After chlorine disinfection, the discharge reaches the standard. The bio-contact oxidation method is a method between the activated sludge method and the biological filter. Its characteristic is that packing is set in the tank, and the bottom aeration is used to oxygenate the sewage and make the sewage in the tank flow. This method ensures that the sewage fully contacts the filler immersed in the sewage and avoids the defect of uneven contact between the sewage and the filler in the bio-contact oxidation tank. Besides, this aeration device is called blast aeration.

3. Domestic-Sewage Treatment Process

Domestic sewage in the mining area enters the domestic-sewage treatment station through the drainage pipe network. Large debris pieces are removed through the coarse grille and then enter the adjustment tank for water quality and volume adjustment. The adjustment tank is equipped with a lifting pump to lift the sewage through the fine grid (for removing the fine suspended matter and floating matter) to the vortex-type grit chamber. The dry sand is collected for external use, and the effluent enters the anaerobic tank.

In the anaerobic tank, under the phosphorus accumulating bacteria, the release of phosphorus is carried out. Concurrently, the easily degradable organic matter is converted, and part of the nitrogen-containing organic matter is ammonified. Then the effluent flows into the anoxic tank by itself. In the anoxic tank, the organic matter continues to undergo hydrolysis and acidification. Parts of the macromolecular organic pollutants are degraded into small molecular substances, and the microorganisms are denitrified. Then the effluent flows into the aerobic tank by itself. In the aerobic tank, the organic matter is further degraded, a large amount of phosphorus is absorbed, part of the mixed liquid is returned to the anoxic tank, and the effluent flows into the radial-flow sedimentation tank by itself.

In the radial-flow sedimentation tank, part of the sedimented sludge is returned to the anaerobic tank. The remaining sludge is discharged into the sludge tank (after filter press and dehydration, the sludge is transported to the aluminum-magnesium alloy supporting power plant for incineration). Then the supernatant enters the intermediate clean tank. The clean water tank's effluent is lifted by the pump and then enters the multi-layer filter for filtration. The filter makes the sewage highly purified. Finally, the effluent enters the disinfection

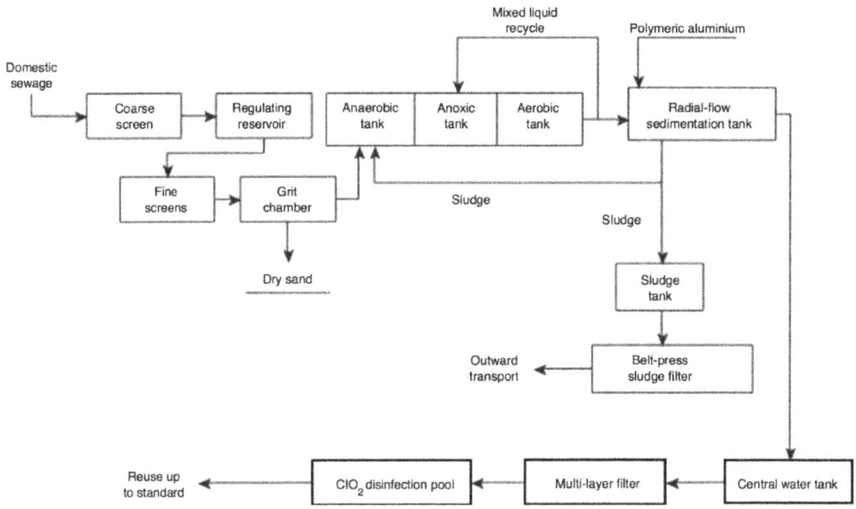

Fig. 1 Domestic-sewage treatment process

tank with residual pressure and overflows to the clean water tank for reuse after disinfection in the disinfection tank.

Figure 1 shows the domestic-sewage treatment process.

Onsite evaluation or supporting materials:

Mining-wastewater treatment design or the volume of domestic-wastewater treatment in a production report.

4 Ore Resources' Comprehensive Utilization

This section applies to industries such as sand and gravel and cement limestone.

43 Ore's comprehensive utilization

Standard scores: 40

Grading description: ★Applicable to the sand and building stone industries: Forty points are rewarded for using the by-products of ore resources such as stone powder and mud powder for processing new building materials, land reclamation, and soil improvement

★Applicable to the cement limestone industry: Forty Points are rewarded for combining the various raw materials' characteristics and the cement-production line's ingredients, realizing resources' classified utilization, high-quality and excellent use, and the comprehensive utilization of high-grade ore and low-grade ore, interlayer, roof, and floor surrounding rocks

Evaluation method: Search data, and check the site
Basis or standard: Production reports (scheduling reports) or other evidentiary materials

[Interpretation] This article requires the comprehensive utilization of waste in the cement and sandstone industries.

1. Sand and Stone, Building Stone Industry

   (1) Stone powder.
   The stone powder is a granular powder with a particle size of less than 0.75 mm produced in the raw ore's production processing into machine-made sand aggregate, and dust collected by environmentally-friendly negative pressure.
   Although the stone powder content is not high, the long-term accumulation will be enormous if it is not used in time. Debris and stone powder's accumulation occupies many land resources and pollutes the environment. Furthermore, the central components of stone powder are the same as the original rock. The stone powder is generally used as sand and gravel enterprises' accessory product. After the stone powder is collected, it should be reasonably stored and comprehensively utilized.
   Calcium carbonate series rocks represented by limestone and marble are produced by processing and producing sand and gravel aggregates. Also, a certain amount of stone powder can be added according to the concrete strength requirements for aggregates. In addition to producing cement, concrete, and mortar, it can be used as a calcium-carbonate powder for desulfurization after deep processing.
   The siliceous series rock represented by granite and tuff is produced by processing and producing sand and gravel aggregates. A certain amount of stone powder can be added according to the demand for aggregates of concrete strength. It can also produce dry-mixed mortar, environmental protection bricks, and new Silica ingredients for wall materials, ceramics, and cement.

   (2) Mud powders.
   The mud precipitated by the wet process's wastewater is dehydrated and dried to form mud powder or mud cake. Its moisture content is below 30%, and its particles are fine and uniform. It can be used to produce new wall materials, land reclamation, and soil improvement, and applied as a cement ingredient according to mud powder or mud cake's chemical composition.

2. Cement Limestone

   (1) Resource hierarchical utilization, high-quality and optimal utilization, and comprehensive utilization.

Resources' classified utilization, high-quality and optimal utilization, and comprehensive utilization refer to the processing of high-quality raw materials into higher-quality products, increasing the added value of products, realizing high-quality and optimal utilization, and using mineral resources. It uses different utilization methods based on cement limestone's different properties and characteristics, market demand, and different raw materials classification.

Mines should reasonably determine the mining scope, deal with the relationship between short-term and long-term production, high-grade ore and low-grade ore, and high-quality raw materials and inferior raw materials, to achieve unified planning, reasonable mining, and comprehensive utilization.

Mines should use computer-related software systems to establish ore deposit models for multi-grade mines, determine the optimal collocation mining plan, and improve mineral resources' utilization rate.

Mines should study the mining conditions and chemical composition of waste rock, determine whether it can be used with ores or used as silicon-alumina raw materials, and develop a comprehensive utilization plan for waste rock.

Mines should study the occurrence and quantity of waste rock and determine whether it can be removed separately, mixed with ore, or separated and stored before being used independently.

(2) Resources' comprehensive utilization.

Low-grade limestone (interlayer rock) can be reasonably matched with high-grade limestone, and the ore will be produced after mixing, and finally meet cement-raw materials' requirements and realize mineral resources' comprehensive utilization.

Soil strips are used as silicon-alumina raw materials or for reclamation; other strips can be used as cement ingredients, sand and gravel aggregates, or other engineering materials to maximize resources' comprehensive utilization and reduce solid-waste emissions.

Onsite evaluation or supporting materials:

Resource development and utilization plan or resource-utilization picture.

5 Solid Waste Disposal and Comprehensive Utilization

This section applies to industries such as sand and gravel and cement limestone.

44 Soil-stripped objects' comprehensive utilization

Standard scores: 40

Grading description: ★Suitable for sand and gravel and building stone industries: Forty points are rewarded for using stripped topsoil or sieved ballast soil and waste rock stacked in dumps for resource utilizations such as new building materials' production, environmental governance, land reclamation, ecological restoration

★Applicable to the cement limestone industry: Twenty points are rewarded for using qualified soil strips as silica-alumina raw materials or reclamation; twenty points for using other strips as cement ingredients, sand and gravel aggregates, or other engineering materials

Evaluation method: Search data, and check the site

Basis or standard: Production reports (scheduling reports) or other evidentiary materials

[Interpretation] This article requires comprehensive utilization of the waste in the cement and sandstone industries.

1. Sand and Gravel Industry

Topsoil is the highest layer of soil. Generally, the Quaternary alluvial layer and rock weathering zone covering the bedrock are collectively called the topsoil layer. Due to the soft topsoil, poor stability, and large changes, the water content is rich, usually 20 to 30 cm at the top. Besides, the topsoil is rich in organic matter and microorganisms.

Topsoil stripping refers to stripping the topsoil covering the upper part and surrounding the open-pit mining ore body.

Ballast soil is a loose covering layer composed of mud and ballast fragments covering the rock.

(1) General requirements.
If the stripped soil contains relatively much gravel, it can be screened to use the resources; if screening is not required, it will be sent to a dedicated dumpsite for centralized storage. After the topsoil is stripped, it is preserved and used for later land reclamation. The stripped topsoil can be used for soil improvement to improve the soil quality in the coverage area.

(2) Reasonable planning.
According to the principle of "who develops, protects; who destroys, who manages," a dumping site should be reasonably planned and set up for stockpiling the stripped topsoil and sieved ballast during the production process. Before mining, the scope, quantity, and extent of each type of land damage should be reasonably estimated based on the land-use status and quality of the mining area and the impact of mining on water and soil damage. Additionally, the Mine Ecological Environment Protection and

Land Reclamation Plan should be formulated to clarify the mine stripping estimated amount, whereabouts, topsoil, and sieved ballast use.

(3) Comprehensive utilization.

It is required to use the stockpiled stripped topsoil and sieved ballast soil for environmental treatment and restoration. The stripped topsoil used for reclamation can effectively protect the surface mellow soil resources from loss and waste and reduce the additional capital investment incurred by external soil adjustment during land reclamation. The use of stripped topsoil for land reclamation has sufficient soil fertility and high crop yields, reduces the extra capital input caused by soil transfer during land reclamation, and can effectively reduce soil erosion.

2. Cement Limestone

Soil strips meeting the requirements should be used as silica-alumina raw materials or for reclamation; other strips can be used as cement ingredients, sand and gravel aggregates, or other engineering materials to maximize resources' comprehensive utilization and reduce solid-waste emission.

Cement production requires lime-based raw materials, silicon-alumina raw materials, iron raw materials, and other raw materials. It can also be used in the production process with high and low-grade ores and various stripped materials to maximize mineral resources' saving and improve resources' comprehensive-utilization rate.

(1) Soil-stripped material.

Soil stripping refers to the surface soil-like materials stripped from the mine, including Quaternary and strong weathered rocks. Its use directions are:

1) Soil strips meeting the requirements, such as clay, are also a raw material for cement production. The soil strips used in mine production can be used as raw materials for cement.
2) The relatively concentrated covering soil can be discarded to waste rock yard or topsoil stacking site for later mine reclamation.

(2) Other stripped objects (rocks).

Other stripped materials (rocks) can be used as a resource based on their chemical composition, such as cement ingredients, sand and gravel aggregates, or other engineering materials.

Various stripped materials' comprehensive utilization reduces solid-waste discharge, thereby reducing land space occupation, maximizes resources' comprehensive utilization, and achieves materials' best use.

Onsite evaluation or supporting materials:

Resource development and utilization plan or resource-utilization picture.

6 Wastewater Treatment and Comprehensive Utilization

This section applies to industries such as sand and gravel and cement limestone.

45 Production-wastewater treatment and utilization
Standard scores: 30

Grading instructions: ① Ten points are rewarded for well-equipping production-wastewater treatment system; ② twenty points for treating wastewater by solid–liquid separation, and effectively recycling clean water

Evaluation method: Search data, and check the site

Basis or standard: Production report (scheduling report) or other evidentiary materials

[Interpretation] This article requires wastewater treatment and comprehensive utilization.

Solid–liquid separation: filtration or precipitation methods are used to remove solids from wastewater.

Water recycling rate refers to the percentage of water reused in total water consumption.

1. Sand and Gravel

   (1) Source of wastewater from sand and gravel production.
   Wastewater generated in the sand and gravel production process essentially comes from spray dust removal, water mist dust suppression, and secondly comes from raw materials and finished products' cleaning and desilting. The sand and gravel processing plant adopts a dry process and high-efficiency bag dust removal. No wastewater is produced. However, if the sand and gravel aggregate products are required to be screened and washed, wastewater will be generated during the washing process. If the wet production process is adopted, wastewater will be generated in the primary process flow. Besides, the suspended matter in the wastewater is chiefly stone powder and sludge.

   (2) Wastewater-treatment system.

      1) The production-wastewater treatment system should be equipped with a solid–liquid separation device to effectively separate the wastewater particulate matter through multi-stage physical and chemical methods.
      2) Wastewater's multi-stage sedimentation and fine sand recycling improve resource utilization; dosing and wastewater's sedimentation reduces wastewater turbidity. Moreover, overflowing-clean water can be recycled.

3) The precipitated sludge achieves clean water's 100% utilization and wastewater's zero discharge by mechanical dehydration, solid–liquid separation, and clean water recycle.

2. Cement Plant

Besides high energy consumption in cement plants, water consumption is also massive. Cement plant production water is often used for cooling, so production water does not require high water quality. The treated reclaimed water can be reused in the circulating pool for industrial production.

(1) Production wastewater's source.

Industrial-production water in cement plants is primarily used in the following aspects:

1) Cooling water for mechanical equipment. The mechanical equipment requiring circulating-cooling water in cement plants includes rotary-kiln tugs, grate cooler oil stations, lubricating oil stations, and reducer-oil-lubrication stations. Its water consumption is large, and the water quality requirements are small, as long as the water temperature is lower than 32 °C. Generally, the reclaimed water after treatment is sufficient.
2) Cooling water for electrical equipment. Electrical equipment requires less circulating cooling water. With technology's advancement, most electrical instruments have been oil-cooled or air-cooled. Only the flame-related TV and a small number of gas analyzers still need circulating cooling water. Its water volume does not exceed $1m^3/h$, and it does not require high water quality. The water temperature is required to be below 32 °C, which can be met by treated reclaimed water.
3) Dust-removal-spray water for limestone, gypsum, and other crushing unloading pits. Dust-removal-spray water for unloading pits in limestone and gypsum crushing workshops essentially uses water spray to reduce dust pollution during unloading. There is no requirement for water quality, as long as there is no sand or other particles in the water to block the nozzle.
4) Laboratory and testing-laboratory test water. Laboratories and testing laboratories demand high-quality test water. It is difficult for industrial water to meet experimental water's requirements. In cement plants, the water used in the laboratory test building is usually the living pool, and the water body meets the drinking water standard. Consider reclaimed water's impact is unnecessary.

(2) Production wastewater treatment.

1) Sewage discharged from the circulating pool. As the circulating cooling water evaporates continuously during the cooling process, the salt concentration in the water is continually increased, easy for scale formation. To ensure qualified cooling water, scale formation's

delay on the condenser's copper tube, and vacuum degree's improvement, it is necessary to add continuously fresh water to the tank. Long-term recycling will inevitably produce a part of effluent.

2) Water-treatment sewage, continuous-boiler sewage, and laboratory wastewater. This wastewater is relatively small, and treatment is qualified (pH is controlled at 6.8–8.2) after the acid–base neutralization. It can be directly used as water spray in the raw material vertical mill, the humidification tower, or the limestone crusher's discharge pit. Besides, it should be no longer recycled.

Onsite evaluation or supporting materials:

Production-wastewater treatment plan.

46 Domestic-sewage disposal

Standard score: Ten

Scoring instructions: ① Four points are rewarded for equipping a domestic sewage treatment system; ② six points for domestic sewage's effective disposal

Evaluation method: Search data, and check the site

Basis or standard: Production reports (scheduling reports) or other evidentiary materials

[Interpretation] This article requires wastewater treatment and comprehensive utilization.

The mine should establish a sewage treatment station to treat domestic sewage centrally. The sewage treatment station needs to have a complete sewage treatment function. See Article 42 for details.

Onsite evaluation or supporting materials:

Domestic-sewage treatment plan.

Chapter 7
Energy Conservation and Emission Reduction

Energy conservation and emission reduction refer to saving material resources and energy resources and reducing the disposal of three wastes, dust and noise emissions. Emission reduction focuses on the energy-saving technology's application, avoiding the sharp increase in energy consumption caused by the unilateral pursuit of emission reduction results, and achieving a balance of economic, social, and environmental benefits.

Energy conservation' evaluation focuses on the level of energy management and the sophistication of the equipment. Emission reduction focuses on the ability to control pollutants.

1 Energy Conservation

47 Energy consumption accounting system for the whole process
Standard score: Five

Instruction: Five points are rewarded for establishing the energy consumption management system for the whole process

Assessment method: Search information

Criteria and standard: Energy consumption system's files or accounts

Explanations: This item is the requirement for mining enterprises to set up the whole process's energy consumption management system.

The whole-process energy consumption accounting system refers to a control system composed of apparatuses, regular monitoring, management systems, responsibility implementation, calculation, and assessment methods. Comprehensive energy consumption refers to the amount of standard coal converted from energy sources,

J. Deng et al., *Interpretation of Green Mine Evaluation Index*,
https://doi.org/10.1007/978-981-16-5433-6_7

such as electric power, gasoline, and diesel oil, in the whole production process from open-pit or underground mining to the exit.

1. Energy consumption system's requirements for the whole process

   (1) Mining enterprises should establish a complete, effective, and documented energy management system and conduct energy consumption accounting during the whole production process to achieve the expected energy consumption targets. We will implement the technical and economic policies of electricity, gas, water, material saving, and environmental protection in construction projects and build a resource-saving and environment-friendly society. Resource-saving and energy utilization rate can be enhanced to the greatest extent by adopting advanced technical measures and management.
   (2) The establishment of the energy consumption accounting system should comply with relevant regulations in the *General Principles to Calculate the Comprehensive Energy Consumption* (GB / T 2589–2008).
   (3) The energy consumption accounting system for the whole process decomposes the annual energy conservation and emission reduction's target, implements it layer by layer, and supervises the earnest implementation with response mechanisms. Energy consumption of mineral resources exploitation and comprehensive energy consumption shall meet the requirements of mine design or cleaner production report, local industrial policies, and industry admittance conditions.
   (4) Energy consumption per unit product refers to a certain amount of energy consumed per unit output or output value. By comparing energy consumption per unit product, we can find out the gap of energy management, ascertain the reasons, analyze the potential, clarify the direction, and put forward countermeasures.
   (5) Energy conservation and emission reduction should satisfy the energy-saving assessment report's requirements.
   (6) Energy management measures. Those enterprises without an energy management system are supposed to establish a series of energy management measures, including energy management, energy statistics, energy metering, and other aspects.

2. Energy conservation's design principles
   Energy conservation should start from design, including several points as follows.

   (1) The technology and device with high efficiency and low energy consumption shall be considered first, and large-scale, automatic electro-hydraulic driving equipment and continuous bulk material transportation equipment shall be widely used.
   (2) Optimize the mining, transportation, processing, storage, and other processes and equipment configurations, simplify the process line, and shorten the process chains' length.

(3) Optimize the mining, transportation, processing, storage, and other process layouts, make full use of the layout space and its natural advantages, and shorten stone materials' transportation distance to avoid reverse transportation.
(4) The wastewater, residues, and dust produced in the mining and aggregate processing should be fully recycled to improve resources' comprehensive utilization rate and use every inch of wastes as much as possible.
(5) According to the different electrical equipment loads and their importance, the appropriate power supply scheme and equipment should be selected to reduce the power consumption in the power supply process, making the power factor above 0.9.
Establish the energy consumption accounting system for the whole process, formulate the energy-saving objectives and assessment system, and implement energy consumption accounting during the whole process.

Onsite evaluation or supporting materials:

Documents of energy consumption accounting system in the whole process.

48 Energy management plans

Standard score: Ten

Instruction: Five points are rewarded for an annual energy management plan; five points for energy conservation indexes settled to subordinate units, departments, or workshops

Assessment method: Looking up information

Criteria and standard: Energy analysis reports

Explanations: This item requires mining enterprises to prepare energy management plans.

Energy management plans.

The energy management plan, considering the needs of the mine, is formulated in accordance with energy conservation, emission, cost reduction, and efficiency management, mainly including work objectives, guiding ideology, overall ideas, main energy conservation measures, and requirements.

Main energy-saving management measures.

(a) Strengthen the target responsibility. Comprehensively considering each unit's production functions, energy conservation and emission reduction targets are differentially implemented to each pollution source unit. The assessment will be reported according to regular inspection results.
(b) Optimize the industrial structure. According to the energy consumption situation of each unit, several secondary units with severe pollution and large energy consumption are eliminated

(c) Promote energy saving and emission reduction equipment. Update and purchase welding fume cleaner and VOCs collection equipment.
(d) Promote energy conservation and emission reduction in critical areas.
(e) Coal mines with high volume gas should adhere to the policy of mining after gas drainage, strengthen comprehensive gas utilization, build gas drainage pump stations and gas power plants, reduce greenhouse gas emissions, and improve enterprise benefits.
(f) Compile energy statistical analysis report. The energy statistical analysis report is significant in expressing energy statistical analysis results with a written report form. Its structure can be generally divided into basic information, achievements and experiences, problems and causes, changes and regularity, development and prospects, suggestions, and measures.

Onsite evaluation or supporting materials:

Energy statistical analysis report.

49 Energy consumption per unit mining product

Standard score: 15

Instruction: Five points deducted for each of the energy, material, and water consumption per unit product failing to meet the specified requirements

Coal, iron, gold, and nonferrous metal ores shall comply with national standards. Supposing there is no national standard or industrial standard for other minerals, the assessment shall be conducted based on the average value of energy consumption and other enterprises' indicators in the recent three years to reflect the energy conservation and consumption reduction progress

Assessment method: Search information

Criteria and standard: Energy consumption account and the quota for energy consumption per unit product in industries

Explanations: This item requires the energy consumption per unit mining product to meet the relevant requirements.

The statistical index of energy consumption per unit product reflects the average energy consumption per unit product (workload). In short, it is the ratio of energy consumption to product output (workload), or a quantitative relationship between product (workload) and energy consumption. Based on this quantitative relationship, we can understand various products' energy consumption levels and estimate the specific product's energy.

1. Coal

    (1) The quota for energy consumption per unit product of existing coal mining.

(1) The quota for energy consumption per unit product of existing coal mining shall meet the requirements of 4.1 in the *Norm of the Energy Consumption per Unit Product of Coal Underground Mining* (GB 29,444–2012): When the conversion coefficient of electricity to standard coal adopts the equivalent value, the limit value of energy consumption per unit product of existing coal underground mining enterprises shall be no greater than 11.8 kgce/t.

(2) The quota for energy consumption per unit product of existing coal open-pit mining should meet the requirements of 4.1 in the *Norm of the Energy Consumption per Unit Product of Coal Open-Pit Mining* (GB 29,445–2012): When the conversion coefficient of electricity to standard coal adopts the equivalent value, the limit value of energy consumption per unit product of existing coal open-pit mining enterprises shall not be greater than 8.2 kgce/t

(2) The quota for energy consumption per unit product of new-built coal mining.

(1) The quota for energy consumption per unit product of new-built coal underground mining shall meet the requirements of 4.2 in the *Norm of the Energy Consumption per Unit Product of Coal Underground Mining* (GB 29,444–2012): When the conversion coefficient of electricity to standard coal adopts the equivalent value, the limit value of energy consumption per unit product of new-built coal underground mining enterprises shall not be greater than 7.0 kgce/t.

(2) The quota for energy consumption per unit product of new-built coal open-pit mining should meet the requirements of 4.2 in the *Norm of the Energy Consumption per Unit Product of Coal Open-Pit Mining* (GB 29,445–2012): When the conversion coefficient of electricity to standard coal adopts the equivalent value, the limit value of energy consumption per unit product of new-built coal open-pit mining enterprises shall not be greater than 6.5 kgce/t.

(3) Advanced value of energy consumption per unit product in coal mining.

(1) The advanced value of energy consumption per unit product of underground coal mining shall meet the requirements of 4.3 in the *Norm of the Energy Consumption per Unit Product of Coal Underground Mining* (GB 29,444–2012): When the conversion coefficient of electricity to standard coal adopts the equivalent value, the advanced value of energy consumption per unit product in coal underground mining enterprises shall not be greater than 3.0 kgce/t through strengthening energy-saving technology and management.

(2) The advanced value of energy consumption per unit product of coal open-pit mining shall meet the requirements of 4.3 in the *Norm of the Energy Consumption per Unit Product of Coal Open-Pit Mining* (GB 29,445–2012): When the conversion coefficient of electricity to standard coal adopts the equivalent value, the advanced value of energy consumption per unit product in coal open-pit mining

enterprises shall not be greater than 3.0 kgce/t through improving energy-saving technology and management.

2. Gold Ore
The quota for energy consumption per unit product of gold mining shall comply with the *Norm of Energy Consumption per Unit Products of Gold Mining* (GB 32,032–2015), and the quota for energy consumption per unit product of gold ore dressing and smelting shall meet the admission value specified in *Norm of energy consumption per unit products for mineral processing and metallurgy of gold* (GB 32,033–2015).

3. Iron Ore
Table 1 shows the limit value of energy consumption per unit product of iron mining.
Table 2 shows the limit value of energy consumption per unit product of iron ore dressing
With the decreased iron ore grade and the increased ore-dressing difficulty, the data in Table 2, not meeting industry development needs, is only for reference.

4. Nonferrous Metal Ores

   (1) Requirements on comprehensive energy consumption index of mining
   See Table 3 for the benchmark (comparable) comprehensive energy consumption index of open-pit mining $P_{0\ \text{open-pit}}$.
   See Table 4 for the benchmark (comparable) comprehensive energy consumption index of open-pit mining $P_{0\ \text{underground}}$.

Table 1 Limit value of energy consumption per unit product of iron mining

Mining method	Mining type	Mine size	Comparable comprehensive consumption/(kgce/t)
Open-pit mining	Existing mines	Above medium (including medium)	≤0.80
		Small	≤1.04
	New-built and extended mines	Above medium (including medium)	≤0.49
		Small	≤0.64
Underground mining	Existing mines	Above medium (including medium)	≤3.60
		Small	≤4.68
	New-built and extended mines	Above medium (including medium)	≤2.60
		Small	≤3.38

Table 2 Limit value of energy consumption per unit product of iron ore dressing

Mining type	Ore dressing type		Comparable comprehensive consumption/(kgce/t)
Existing mines	Low-intensity magnetic separation		≤4.1
	Combined dressing		≤5.7
	Roasting	Shaft furnace	≤48.5
		Rotary kiln	≤54.3
New-built and extended mines	Low-intensity magnetic separation		≤3.3
	Combined dressing		≤4.2
	Roasting	Shaft furnace	≤45.6
		Rotary kiln	≤51.8

Table 3 Benchmark (comparable) comprehensive energy consumption index of open-pit mining $P_{0\ \text{open-pit}}$

Design scale	$P_{0\ \text{open-pit}}$/[kg standard coal/t_{mine} (kW·h/t_{mine})]		
	Level 1	Level 2	Level 3
Large	0.95(7.73)	1.05(8.54)	1.15(9.36)
Medium	1.25(10.17)	1.35(10.98)	1.50(12.21)
Small	1.63(13.26)	1.75(14.24)	1.95(15.87)

Table 4 Benchmark (comparable) comprehensive energy consumption index of open-pit mining $P_{0\ \text{underground}}$

Design scale	$P_{0\ \text{underground}}$/[kg standard coal/$t_{mine}$ (kW·h/t_{mine})]		
	Level 1	Level 2	Level 3
Large	1.84(15)	2.21(18)	2.70(22)
Medium	2.21(18)	2.70(22)	3.20(26)
Small	2.70(22)	3.20(26)	3.81(31)

(2) Comprehensive energy consumption index of mineral processing
See Table 5 for the comprehensive energy consumption index of mineral processing.

5. If there is no national standard or industrial standard for other minerals, the assessment shall be conducted based on the average value of energy consumption and other enterprises' indicators in the recent three years to reflect the energy conservation and consumption reduction's progress.

Table 5 Comprehensive energy consumption index of mineral processing

Metal	Ore type	Comprehensive energy consumption of mineral processing(kW·h/$t_{raw\ ore}$)		
		Level 1	Level 2	Level 3
Cuprum	Sulfide ore	≤22	<22 ~ 27	< 27 ~ 34
	Mixed ore Oxide ore	≤26	<26 ~ 32	<32 ~ 41
Molybdenum	Sulfide ore	≤20	<20 ~ 25	<25 ~ 31
Nickel	Sulfide ore	≤42	<42 ~ 48	<48 ~ 55
Lead–Zinc	Sulfide ore	≤29	<29 ~ 37	<37 ~ 46
	Mixed ore Oxide ore	≤35	<35 ~ 44	<44 ~ 55
Stibium	Sulfide ore	≤17	<17 ~ 21	<21 ~ 26
Mercury	Cinnabar	≤26	<26 ~ 31	<31 ~ 37
Wolfram	Wolframite	≤8	<8 ~ 11	<11 ~ 15
	Scheelite Mixed ore	≤23	<23 ~ 30	<30 ~ 39
Tin	Sulfide ore	≤38	<38 ~ 48	<48 ~ 60
	Oxide ore	≤24	<24 ~ 30	<30 ~ 38
	Placer	≤13	<13 ~ 16	<16 ~ 20

Note The comprehensive energy consumption of mineral processes includes crushing, screening, grinding, separation, concentrate dewatering, and other production processes, excluding tailings, water supply, heating, and other auxiliary processes

Onsite evaluation or supporting materials:

(1) Energy conservation and consumption reduction report or energy conservation and emission reduction report.
(2) Viewing the original account of energy, material, and water consumption in the whole process of mine.

50 Energy management system certification

Standard score: Five

Instruction: Five points for energy management system certification

Assessment method: Checking certificates

Criteria and standard: Energy management system certificate

Explanations: This item encourages mining enterprises to obtain energy management system certification.

Based on the system's whole process and principles of system management, the energy management system (*Energy management systems—Requirements* (GB/T

23,331—2012)) is used to establish a complete, effective, and documented system in the organizations by implementing a complete set of standards and specifications. This system focuses on establishing and implementing process controls to optimize organizations' activities, processes, and elements.

Through routine energy-saving monitoring, energy audit, energy efficiency benchmarking, internal audit, organizations' energy consumption measurement, test and energy balance statistics, management review, self-evaluation, energy-saving technical renovation, energy-saving assessment, and other measures, we can improve the energy management system to achieve energy management policies, expected energy consumption, or utilization objectives.

According to PDCA management mode, the energy management system promotes energy management, i.e., the continuous improvement model based on plan-do-check-action. Plan: Implement energy assessment, define energy benchmarks and energy performance parameters, set energy targets, indicators, and energy management implementation plans to ensure the organization improves energy performance following its energy policy. Implementation: Conduct an energy management plan. Check: Monitor and measure key operating characteristics and processes, assess energy performance according to energy guidelines and objectives, and report results. Action: Take action to improve energy performance and energy management systems continuously.

Establishing, implementing, and maintaining an environmental management system reduces or eliminates hazards to human health and the environment by continuously improving environmental management levels, reducing and avoiding the pollutants' generation and emission in the process of production, service, and use.

Onsite evaluation or supporting materials:

Energy management system certification.

2 Exhaust Emissions

51 Main dust processing sites' list

Standard score: Five

Instruction: Mines have transparent dust-producing workplaces such as mining, transportation, mineral processing, and a list of dust concentrations at their posts

Assessment method: Check the site

Criteria and standard: Relevant measures for dust prevention in enterprises

Explanations: This item requires enterprises to sort out dust production points and dust concentration at their posts.

This article briefly describes mines' dust production points from three aspects: dust control points of sandstone aggregate production lines, dust control points of non-ferrous metal production lines, and dust control points of black metal production lines.

1. Dust Removal Points of Sandstone Aggregate Production Lines

   (1) Raw ore bin's dust removal points.
   Raw ore's feeding port.
   (2) Dust removal points in coarse crushing.
      (a) Receiving point on the upper part of the feeder
      (b) Feeding port on the upper part of crusher
      (c) Crusher discharge's dropping point on the belt
   (3) Dust removal points in the semi-finished product warehouse.
      (a) Incoming material belt's head
      (b) Incoming material belt's head returns to the roller
      (c) The belt's dropping point in the semi-finished product warehouse
      (d) Feeder with the bottom discharge hole in the semi-finished product warehouse
      (e) The feeder unloads the material to the discharge belt
   (4) Dust removal points in desilting screening.
      (a) Incoming material belt's head returns to the roller
      (b) Desilting screen surface
      (c) Screen overflow drops on the conveyor belt
      (d) Screen underflow drops on the conveyor belt
   (5) Dust removal points of waste residue.
      (a) Incoming material belt's head returns to the roller
      (b) Waste residue's incoming belt unloads
      (c) The dropping point of waste residue's belt
   (6) Dust removal points in medium crushing.
      (a) Moderating bunker's upper feeding port of the medium crusher
      (b) Feeding machine's surface
      (c) Upper feeding port of the medium crusher
      (d) Medium crusher discharge's bottom dropping point on the conveyor belt
   (7) Dust removal points in fine crushing.
      (a) Moderating bunker's upper feeding port of the fine crusher
      (b) Dust producing points on the feeding machine surface
      (c) Upper feeding port of the fine crusher
      (d) Fine crusher discharge's bottom dropping point on the conveyor belt
   (8) Dust removal points in primary screening.
      (a) Incoming material belt's head

   (b) Moderating bunker's upper feeding port in primary screening
   (c) Dust producing points on the feeding machine surface
   (d) Primary vibrating screen (multilayer screen) surface
   (e) Primary vibrating screen's overflow drops on the conveyor belt
   (f) Primary vibrating screen's medium layer drops on the conveyor belt
   (g) Primary vibrating screen's underflow drops on the conveyor belt

(9) Dust removal points in secondary screening.
   (a) Incoming material belt's head
   (b) Moderating bunker's upper feeding port in secondary screening
   (c) Feeding machine's surface
   (d) Secondary vibrating screen's (multilayer screen) surface
   (e) Secondary vibrating screen's overflow drops on the conveyor belt
   (f) Secondary vibrating screen's medium layer drops on the conveyor belt
   (g) Secondary vibrating screen's underflow drops on the conveyor belt

(10) Dust removal points in shaping.
   (a) Moderating bunker's upper feeding port of the shaper
   (b) Dust producing points on the feeding machine surface
   (c) Upper feeding port of the shaper (Vertical shaft impact crusher)
   (d) Shaper (Vertical shaft impact crusher) discharge's bottom dropping point on the conveyor belt

(11) Dust removal points in shaping section's screening.
   (a) Incoming material belt's head
   (b) Moderating bunker's upper feeding port in shaping screening
   (c) Dust producing points on the feeding machine surface
   (d) Vibrating screen surface
   (e) Vibrating screen's medium layer drops on the conveyor belt
   (f) Vibrating screen's underflow drops on the conveyor belt

(12) Dust removal points in the finished aggregate warehouse.
   (a) Top feeding port of the finished product warehouse
   (b) Belt transferring point at the top of the warehouse
   (c) Finished product belt's head returns
   (d) Finished product loading

(13) Dust removal points in stone powder tanks.
   (a) Discharge gas from the top of the tank
   (b) Eliminate the powder at the bottom of the tank

2. Dust Removal Points of the Non-Ferrous Metal Production Line

(1) Raw ore bin's dust removal points.
   Raw ore's feeding port.

(2) Dust removal points in coarse crushing.
   (a) Receiving point on the upper part of the feeder
   (b) Feeding port on the upper part of crusher

   (c) Crusher discharge's dropping point on the belt
(3) Dust removal points in medium crushing.
   (a) Storage bunker's upper feeding port of the medium crusher
   (b) Feeding machine surface
   (c) Upper feeding port of the medium crusher
   (d) Medium crusher discharge's bottom dropping point on the conveyor belt
(4) Dust removal points in fine crushing.
   (a) Storage bunker's upper feeding port of the fine crusher
   (b) Dust producing points on the feeding machine surface
   (c) Upper feeding port of the fine crusher
   (d) Fine crusher discharge's bottom dropping point on the conveyor belt
(5) Examine dust removal points in screening.
   (a) Incoming material belt's head
   (b) Storage bunker's upper feeding port
   (c) Feeding machine surface
   (d) Vibrating screen surface
   (e) Vibrating screen's unqualified materials from overflow drop on the conveyor belt
   (f) Vibrating screen underflow's finished product drops on the conveyor belt
(6) Dust removal points in the transferring station.
   (a) Incoming material belt's head
   (b) Incoming material belt transfers and unload to the downstream belt
(7) Dust removal points at the top of the finished product warehouse.
   (a) Incoming material belt's head
   (b) Incoming material belt's dropping point on the moving material belt.
   (c) Moving belt drops the material at the finished product warehouse
(8) Dust removal points at the bottom of the finished product warehouse.
   (a) Feeding machine's surface
   (b) Feeding machine unloads the material to the transferring belt
   (c) Transferring belt unloads the material to the output long belt
   (d) Transferring belt's head
   (e) Output long belt's head

3. Dust Removal Points of the Ferrous Metal Production Line

(1) Raw ore bin's dust removal points.
   Raw ore's feeding port.
(2) Dust removal points in coarse crushing.
   (a) Receiving point on the upper part of the feeder
   (b) Feeding port on the upper part of crusher
   (c) Crusher discharge's dropping point on the belt
(3) Dust removal points in medium crushing.

(a) Storage bunker's upper feeding port of the medium crusher
(b) Feeding machine surface
(c) Upper feeding port of the medium crusher
(d) Medium crusher discharge's bottom dropping point on the conveyor belt.

(4) Dust removal points in fine crushing.
(a) Storage bunker's upper feeding port of the fine crusher
(b) Dust producing points on the feeding machine surface
(c) Upper feeding port of the fine crusher
(d) Fine crusher discharge's bottom dropping point on the conveyor belt

(5) Examine dust removal points in the screening.
(a) Incoming material belt's head
(b) Storage bunker's upper feeding port
(c) Feeding machine surface
(d) Vibrating screen surface
(e) Vibrating screen's unqualified materials from overflow drop on the conveyor belt
(f) Vibrating screen underflow's finished product drops on the conveyor belt

(6) Dust removal points in dry magnetic separation.
(a) Receiving point of the dry magnetic separator
(b) Dry magnetic separator discharge's dropping point on the concentrate belt
(c) Dry magnetic separator discharge's dropping point on the waste ore belt
(d) The waste ore belt drops materials to scavenging's (dry magnetic separator) feeding port
(e) Scavenging (dry magnetic separator) discharge's dropping point on the concentrate belt
(f) Scavenging (dry magnetic separator) discharge's dropping point on the waste ore belt

(7) Dust removal points in the transferring station.
(a) Incoming material belt's head
(b) Incoming material belt transfers and unload to the downstream belt

(8) Dust removal points at the top of the concentrate warehouse.
(a) Concentrate belt's head
(b) Concentrate belt's head returns
(c) Concentrate belt's dropping point on the moving material belt.
(d) Moving belt drops the material at the concentrate warehouse

(9) Dust removal points at the bottom of the finished product warehouse.
(a) Feeding machine's surface
(b) Feeding machine unloads the material to the transferring belt
(c) Transferring belt unloads the material to the output long belt
(d) Transferring belt's head

(e) Output long belt's head

(10) Dust removal points of waste ore pile.

(a) (a) Waste ore belt's head.

(b) (b) Waste ore belt's head returns.

(c) (c) Waste ore belt drops materials at the waste ore pile.

4. Common Mine Posts' Dust Concentration Requirements
Common mine posts' dust concentration requirements refer to *Occupational Exposure Limits for Hazardous Agents in the Workplace Part 1: Chemical Hazardous Factors* (GBZ 2.1–2007). See Table 6 for details.
Explanations for Table 6

(1) There are two kinds of dust, total dust and respirable dust.

(2) Total dust is all dust that can enter the whole respiratory tract (nose, pharynx, larynx, thoracic bronchus, bronchioles, and alveoli). Technically, it refers to all dust measured in the breathing zone by the standard method with the total dust sampler, or sampled by the standard dust measurement method with a 40-mm-diameter filter.

(3) Respirable dust is the particulate that can enter the human alveoli collected according to the standard measurement method of respirable dust. Its aerodynamic diameter is less than 7.07 μ m, and the sampling efficiency of 5 μ m particles is 50%.

5. Dust List
Table 7 shows the dust list.

Onsite evaluation or supporting materials:

Dust production post's concentration list.

52 Dust emission in production

Standard score: 15

Instruction: (1) Use rock drilling and dust-collecting integrated drilling machine or wet drilling. (2) Spray water in blasting operations. (3) Fix dust production point, add dust removal equipment, and keep enough negative pressure with production equipment to reduce dust in mining and dressing. Three points deducted for each non-conformance in key links or locations such as drilling, blasting, rock (ore) crushing (pulverizing), screening, conveying, and dosing

Assessment method: Check the site, and employees randomly

Criteria and standard: Blasting involved mines should provide a special dust reduction scheme, while other open-pit mines in loose rock stratum are not required

Table 6 Common mine posts' dust concentration requirements

Sequence number	Name	Chemical Abstract Service Number (CASNo.)	Permissible concentration–Time Weighted Average (PC-TWA)/(mg/m^3)		Critical adverse health effects	Remarks
			Total dust	Respirable dust		
1	Dolomite	–	8	4	Pneumoconiosis	–
2	Marble	1317–65-3	8	4	Eyes and skin irritation; Pneumoconiosis	–
3	Wollastonite	13,983–17-0	5	–	–	–
4	Diatomite (Free silica content < 10%)	61,790–53-2	6	–	Pneumoconiosis	–
5	Calcium carbonate	1317–65-3	8	4	Eyes and skin irritation; Pneumoconiosis	–
6	10% ≤ Free silica content ≤ 50%	14,808–60-7	1	0.7	Silicosis	G1 crystalline form
	50% ≤ Free silica content ≤ 80%		0.7	0.3		
	Free silica content > 80%		0.5	*0.2*		
7	Rare earth (Free silica content < 10%)	–	2.5	–	Rare earth pneumoconiosis; Skin irritation	–
8	Fluorite	–	1	0.7	Silicosis	–
9	Mica	12,001–26-2	2	1.5	Mica pneumoconiosis	–

(continued)

Table 6 (continued)

Sequence number	Name	Chemical Abstract Service Number (CASNo.)	Permissible concentration–Time Weighted Average (PC-TWA)/(mg/m^3)		Critical adverse health effects	Remarks
			Total dust	Respirable dust		
10	Perlite	93,763–70-3	8	4	Eyes, skin, and upper respiratory tract irritation	–
11	Vermiculite	–	3	–	Eyes and skin irritation	–
12	Barite	7727–43-7	5	–	Eyes irritation; Pneumoconiosis	–
13	Other powders	–	8	–	–	–

Table 7 Dust list example

Workspace	Main device	Total Dust/ (mg/m^3)	Respirable Dust/ (mg/m^3)
Rock drilling post and surrounding workplace	DTH drill	8	4
Excavator	Loader	8	4
Crushing workshop	Crusher, vibrating screen	8	8

Explanations: This article is about the requirement for dust removal measures in production.

The dust concentration's qualified rate in underground mines is only 40–60%, and that in open-pit mines is only 70–80%. Organized emissions prove to be more effective in raising the dust concentration's qualified rate than non-organized emissions.

1. Main sources of productive dust

    (1) Solid particles are formed by mechanical impact, grinding and rolling, and suspended in the air by airflow, such as metal grinding, cutting, drilling, blasting, crushing, grinding powder, agricultural and forestry product processing.
    (2) When a substance is heated in the air, its vapor will condense, oxidize, and transfer into smoke and dust, such as metal smelting, welding, and casting.
    (3) Smoke from incomplete combustion of organic matter, such as wood, oil, and coal.

2. Dust Production in Mining
    Large quantities of dust are produced in the mining process, including perforation, blasting, secondary crushing, shovel loading, truck transportation, truck unloading, crushing, loader's level working surface, and dump. Open-pit mine stopes always have many dust production points with large dust amounts, high concentration in the air, and high dispersion.

    (1) Dust produced in drilling rig perforation.
    More dust will be produced without any measures when rocks are crushed into powder during perforation.
    (2) Dust produced in blasting operation.
    During the blasting operation, dust is formed after ore and rock are crushed by the considerable pressure of charge blasting. Blasting produces a maximum amount of dust, while the high-concentration dust will not exist for long.
    (3) Dust produced in shovel-loading operation.

Some of the dust is settled on the mine rock surface, and others are secondary dust lifted by vibration. Besides, a large amount of dust is generated due to the drop when the bucket is unloading to the truck.

(4) Dust produced in crushing operation.
A great deal of dust is produced during ore crushing. With no effective measures in this process, the dust hazard will have an acute impact on employees and cause severe dust pollution.

3. Measures
For non-organized dust raising in mining, the wet operation shall be adopted for three dust-producing points: shoveling, loading, and transporting. Water spraying, covering, and dust suppression measures shall be introduced into a dust-generating surface such as exposed mining faces and waste rock yards. Finally, cover soil and plant suitable vegetation to restore the ecology.

(1) Dust prevention in drilling. Wet drilling with high dust removal efficiency is much more preferred. When wet drilling conditions are unavailable, dry dust-collecting drilling is adopted, which can be divided into two types: using dust collectors at the chisel opening or dust extraction at the bottom of the hole. Foam can also be used for dust removal if large quantities of water are not permitted. Foams will be pressed into the bottom of the hole or sprayed on the orifice, or dust-laden air is introduced into the foam cleaner for purification.
(2) Dust prevention in blasting. When blasting ore, there is the largest amount of dust, and the dust production time is concentrated. The main dust-proof measures are spraying water and water-seal blasting. Blastwave is often used to start sprayer, purify water curtain, and water–air ejector.
(3) Dust prevention in shipment. Several sectors may produce a large amount of dust, including mine stacking, rock loader, loader, scraper, mine truck, belt conveyor, slide ore drawing, and loading. Spraying water, closed air extraction, and other measures can be used to prevent dust emission.
(4) Dust prevention in crushing. Apart from air-tight equipment, wet dedusting and mechanical dedusting measures can be adopted to reduce the dust generated by various crushing equipment. Bag-type or wet-type dust collector is used as the cleaning equipment for mechanical dedusting. A wet-type dust collector is used for dust removal in an underground crushing chamber. When the dust removal system's exhaust air can only be discharged locally, rather than to the return tunnel or ground, the exhaust air shall pass through the high-efficiency filter.

4. Installation of High-Efficiency Dust Removal Device in the Production System

(1) Crushing, screening, and feeding systems shall be in a fully sealed workshop. Bag type dust collector, sprinkler, and centralized negative pressure dust collector are used for systems' comprehensive management.

(2) In the process of crushing (pulverizing), screening, conveying, and proportioning, high-efficiency dust removal devices including high-efficiency bag dust collector, dust suppressor, guide chute should be installed at the places where dust is continuously generated. Configurations of efficient dust removal equipment for sand production and processing refer to Table 8

(3) Material transportation in production should use closed belts and corridors, and materials should be stored in closed places.

5. Dust Protection Requirements in Storage and Transportation
For lifting dust during mining and transportation, vehicle washing facilities shall be set up at the factory (yard) to flush and remove mud; ore loading shall be no higher than the cabin and canvas to control causing dust and pollutions during ore transportation. Personnel shall be arranged to sweep the roads and spray water in time to control the dust in the villages they pass through. Drive carefully and slowly through villages to reduce vehicle bumps and ore spills.
For coal mines: (1) Coal storage yard shall be closed entirely with water spraying and dust suppression facilities. (2) Waste rock's temporary turnaround area should be a canopy roof to form a semi-closed area, and dust suppression measures like spraying water should be adopted. (3) The air separation system shall be equipped with a bag-type dust collector. The dust concentration

Table 8 Configurations of efficient dust removal equipment for sand production and processing

Dust-generating processes	Dust points and measures	Dust removal equipment
Coarse crushing	The truck's unloading port is closed on all sides, and a bag-type dust collector is installed for negative pressure dust collection with water mist and dry fog; the feeding port of feeder and crusher is equipped with a bag dust collector with water mist	Dust suppressor, bag filter
Screening	Partially closed with bag-type dust collector, supplemented by water mist	Closed environment, bag-type dust collector
Medium and fine crushing	Partially closed with bag-type dust collector, supplemented by water mist	Closed environment, dust suppressor
Transport belt and belt's transfer point	Guide chute at belt's transfer points and feeding inlet, dry fog at the tail, partially closed with bag-type dust collector	Closed environment, guide chute
Storage yard	Fully closed, installing inductive high-efficiency water-spray dust-suppression system	Closed environment, fog dust suppression

and removal efficiency at the exhaust cylinder's outlet shall meet the relevant requirements of the *Emission Standard for Pollutants from Coal Industry* (GB 20,426–2006).

6. Limit of Atmospheric Pollutant Emission Concentration
   The latest emission limits for atmospheric pollutants are in Table 9

Some regions have formulated local standards. Table 10 shows air pollutants' emission concentration limits in the *Regional and Integrated Emission Standard of Air Pollutants* (DB37_ 2376—2019) of Shandong Province.

Coalbed methane emission shall meet the standard for gas discharge in *Emission Standard of Coalbed Methane/Coal Mine Gas* (GB 21,522–2008).

Onsite evaluation or supporting materials:

Dust removal measures at critical locations such as drilling, blasting, rock (ore) crushing (pulverizing), screening, conveying, and dosing.

53 Dust emission in ground transportation

Standard score: 15

Instructions: Avoiding dust flying can be achieved from the following aspects: Setting water spraying or inductive spraying facilities along the transportation road or equipping sprinklers to spray water and dust regularly; Spraying or water for ground transport vehicles and devices' dust removal; Using sealed vehicles for shipping. Three points will be deducted for each non-conformance

Assessment method: Check the site

Explanations: This article is about the requirement for dust generation in ground transportation.

Production and transportation work transports the ore and rock of the open pit to the unloading point (or concentrator) and the dump separately and transports the production personnel, equipment, and materials to the mining site at the same time.

1. Reasons for dust production in transportations
   One important dust source in mines comes from the dust lifted by vehicles. Besides, dust deposited on the road moves irregularly because of the extrusion, vibration, and airflow generated by vehicles, and finally forms secondary dust. Dust is also generated by rock friction when dumping.
2. Dust protection measures
   Wash the vehicles, spray water before loading and spray water mist at unloading sites; reinforce pavement protection, seal the shipping vehicles, and install spraying dust controllers or sprinklers.

Table 9 Latest emission limits for atmospheric pollutants

Number	Pollutant	Upper limit for emission concentration/(mg/m^3)	Analysis method	Method source
1	Sulfur dioxide	500	Formaldehyde absorbing-pararosaniline spectrophotometry	(HJ 482—2009)
			Iodometry	HJ/T 56
			Constant potential electrolysis	HJ/T 57
2	Nitrogen oxide	120	Saltzman	GB/T 15,436
			UV-Spectrophotometry	HJ/T 42
			N-(1-naphthyl)-ethylenediamine dihydrochloride spectrophotometric method	HJ/T 43
			Constant potential electrolysis	A
3	Hydrogen chloride	100	Ion chromatography	A
			Mercury Thiocyanate Spectrophotometry	HJ/T 27
4	Chlorine	65	Methyl orange spectrophotometry	HJ/T 30
5	Methanol	190	Gas chromatography	HJ/T 33
6	Carbon monoxide	1000	Non-dispersive infrared absorption method	GB 9801
			Constant potential electrolysis	A
			Non-dispersive infrared absorption method	HJ/T 44

Table 10 Emission concentration limits of air pollutants

Pollutants	Core control area	Key control area	Normal control area
Particulate matter	5	10	20
Carbon dioxide	35	50	100
Nitrogen oxides (as nitrogen dioxide)	50	100	200

Onsite evaluation or supporting materials:

Take measures in transportations.

54 Dust emissions in storage places

Standard score: Ten

Instructions: (1) Five points are rewarded for dust prevention facilities in waste rock or ore turnover yards and storage places. (2) Five points for achieving dust prevention

Assessment methods: Search information and checking the sites

Criteria and standard: Relevant measures for dust prevention in enterprises

Explanations: This article is about the requirement for dust emissions in storage places.

1. Temporary waste stone or ore stacking yards' impact on the atmospheric environment

   Waste rocks are discarded from stripping surface rocks of open-pit mines, industrial squares and road construction in the mine area, well construction, and mining. Waste rocks' temporary piling occupies land, pollutes soil and water, results in landslides, debris flow, and other disasters, and affects the surrounding atmospheric environment.

   The atmospheric pollution caused by waste rock piling is mainly manifested in dust pollution and exhaust gas pollution.

   (1) Dust pollution

   Micro dust entering the human lung can persist for a long time and have a chronic effect on respiration. Dust falling on vegetation and crop surfaces hinders their photosynthesis. Humans and animals will have problems in breathing in high dust concentration environments. Tailings and waste rocks are the most important sources of dust pollution. The fine particles of tailings and stones aggravate the mining area's pollution, especially in dry and windy seasons.

(2) Waste gas pollution caused by coal gangue piles' spontaneous combustion
In coal mine areas, pyrite in gangue is oxidized by air after sunshine, rain, weathering, and denudation. Heat impels coal in gangue to spontaneous combustion, releasing hydrogen sulfide, sulfur dioxide, carbon monoxide, and carbon dioxide.

2. Measures for air pollution prevention and control in the temporary storage of waste rocks or ores

(1) Dust prevention measures in the waste rock yard
The amount of dust in the waste rock yard is related to the ore's density, grain size, humidity, and wind speed. Water spraying is the main method for dust control. It is generally suitable to wash hard rock and large waste rock with a water gun, and a water cooler is suitable for soft and dusty rock and soil.
Besides, spraying the covering agent on the surface of the waste rock heap is also an effective dust suppression method. Due to the bonding force, the covering agent and the waste rock would penetrate each other. Under the chemical bond force and physical adsorption, a thin layer of hard shell is formed on the surface of waste rock to prevent dust caused by wind, rain, and sunshine. The covering agent material is mainly composed of tar, acid tar, anticorrosive oil, polyvinyl acetate, emulsifier, and water.
(2) Measures for coal gangue piles' spontaneous combustion
The prevention of coal gangue piles' spontaneous combustion mainly starts from the research in different stacking methods. Many countries stack coal gangue by layered compaction method and plant trees on the side slope at the same time. The spontaneous combustion problem of coal gangue piles can be solved by air isolation, which directly depends on the compaction degree of coal gangue. For the coal gangue with high sulfur content, lime slurry and clay slurry should be used to fill the voids to isolate air and inhibit spontaneous combustion. For the coal gangue with high carbon content, the coal should be recycled or used as fuel in a fluidized bed furnace. However, the fundamental way to reduce and avoid coal gangue piles' spontaneous combustion is to enhance coal gangue resource utilization.
(3) Others
Other enterprises also use the plastic dust-proof net to control the dust in the open air.

Onsite evaluation or supporting materials:

(1) Dust prevention measures
(2) Relevant photos

55 Other exhaust emissions

Standard score: Ten

Instructions: Ten points are rewarded for qualified exhaust purification systems for the industrial waste gas contains toxic and harmful substances (such as SO_2 and NO_x) except dust in the mining and dressing

Assessment methods: Look up information

Criteria and standard: Monitoring report or test data

Explanations: This article is the requirement for exhaust emissions.

Mine waste gas pollution refers to the pollution caused by waste gas in the mining process. Mine waste gas includes blasting smoke and dust, gas from sulfide ore's oxidation, spontaneous combustion and hydrolysis, and harmful gas from radioactive elements in the uranium deposit.

The enterprise shall provide the test report and relevant testing items' qualification certification issued by the qualified testing department.

1. Detection requirements of waste gas (dust) concentration

    (1) Refer to Table 1 "Allowable chemical substances concentration in the air of workplace" limit value of *Occupational Exposure Limits for Hazardous Agents in the Workplace Part 1: Chemical Hazardous Factors* (GBZ 2.1–2007).

    (2) If there is no industrial pollution emission standard for non-metallic mines, the emission concentration of exhaust gas (dust) should meet the secondary standard in Table 2 of the *Integrated Emission Standard of Air Pollutants* (GB 16,297–1996). The waste gas (dust) of coal mine and building-use limestone mine shall comply with the emission standard of *Emission Standard for Pollutants from the Coal Industry* (GB 20,426–2006) and *Emission standard of air pollutants for cement industry* (GB 4915–2013) separately.

    (3) The waste gas emission from iron ore mining and dressing shall comply with the *Emission Standard of Pollutants for Mining And Mineral Processing Industry* (GB 28,661—2012).

    (4) If there is no industrial pollution emission standard for nonferrous metal mines, the emission concentration of waste gas (dust) shall comply with should meet the secondary standard in Table 2 of *Integrated Emission Standard of Air Pollutants* (GB 16,297–1996).

    (5) The waste gas emission in nonferrous metal mines should comply with the following standards: *Emission Standard of Pollutants for Aluminum Industry* (GB 25,465—2010), *Emission Standard of Pollutants for Lead and Zinc Industry* (GB 25,466—2010), *Emission Standard of Pollutants for Copper, Nickel, Cobalt Industry* (GB 25,467—2010), *Emission Standards of Pollutants for Stannum, Antimony and Mercury Industries* (GB

30,770—2014), *Discharge Standard of Pollutants for Vanadium Industry* (GB 26,452—2011), *Emission Standard of Pollutants for Magnesium and Titanium Industry* (GB 25,468—2010).

2. Waste gas standard emissions
   (1) Install the waste-gas pollutant-detection notice board and test-result display screen at suitable places in the plant area.
   (2) Tail gas test report of transport vehicles shall be provided, and the exhaust emission shall meet the standard.

Onsite evaluation or supporting materials:

Test reports from a qualified testing department provided by enterprises.

3 Wastewater Discharge

56 Domestic sewage discharge

Standard score: Ten

Instructions: Ten points are rewarded for domestic sewage after treatment meeting the discharge standards or directly discharging into the municipal sewage pipe network

Assessment methods: Search information and check the site

Criteria and standard: Acceptance data from environmental protection facilities such as sewage station

Explanations: This article is the requirement for domestic sewage discharge.

Establish sewage treatment stations to treat domestic sewage centrally. Sewage treatment stations shall have a complete sewage treatment function.

1. Standard classifications of sewage treatment
   According to water quality analysis, industrial wastewater can be divided into drainage without exceeding heavy metal ions and drainage with exceeding heavy metal ions. The judgment criteria of whether sewage treatment is necessary include pollutant discharge standard for local water systems and *Integrated Wastewater Discharge Standard* (GB 8978–1996). Generally, some drainage indexes of local discharge standards are higher than those of *Integrated Wastewater Discharge Standard* (GB 8978–1996); in the absence of local discharge standards, *Integrated Wastewater Discharge Standard* (GB 8978–1996) shall be strictly implemented. Pollutants discharged into water are classified into two categories according to their properties and control methods in the *Integrated Wastewater Discharge Standard* (GB 8978–1996).

Article 4.2.1.1 in the *Integrated Wastewater Discharge Standard* (GB 8978–1996) stipulates that the first group of pollutants, regardless of industry, sewage discharge mode, and functional category of the accepted water body, shall be sampled at the discharge outlet of the workshop or workshop treatment facility. Its maximum permissible discharge concentration must meet the standard requirements (tailings dam outlet in mining industry shall not be regarded as workshop discharge outlet).

The first group of pollutants has 13 indexes. Table 11 shows the maximum permissible emission concentration.

Article 4.2.1.2 in the *Integrated Wastewater Discharge Standard* (GB 8978–1996) stipulates that the second group of pollutants shall be sampled at the discharge unit's discharge outlet. Its maximum permissible discharge concentration must meet the standard requirements.

There are 26 categories of pollutants in the second group, involving 56 indexes (See Table 12). Discharged substances standards are related to the criteria in the region. In non-ferrous mines, the main discharge indexes include pH, suspended substances, BOD5, chemical oxygen demand, total cyanide, ammonia nitrogen, fluoride, total copper, and total zinc.

2. Sewage treatment standard classification
   Integrated wastewater discharge standard (GB 8978–1996) has the following regulations.
   4.1.1 Sewage discharged into Class III waters (excluding designated protected areas and swimming areas) in the *Environmental Quality Standards for Surface Water* (GB 3838–2002) and Class II waters in the *Sea Water Quality Standard* (GB 3097–1997) shall be meet the first-level standards.

Table 11 First group of pollutants

Number	Pollutant	The maximum permissible emission concentration
1	Total mercury	0.05 mg/L
2	Alkyl mercury	N.d
3	Total cadmium	0.1 mg/L
4	Total chromium	1.5 mg/L
5	Hexavalent chromium	0.5 mg/L
6	Total arsenic	0.5 mg/L
7	Total lead	1.0 mg/L
8	Total nickel	1.0 mg/L
9	Benzo(a)pyrene	0.00003 mg/L
10	Total beryllium	0.005 mg/L
11	Total silver	0.5 mg/L
12	Total alpha radioactivity	1 Bq/L
13	Total beta radioactivity	10 Bq/L

Table 12 Second group of pollutants

Number	Pollutant	Application	First-class standard	Second-class standard	Third-class standard
1	pH	All discharge units	6 ~ 9	6 ~ 9	6 ~ 9
2	Chroma (Dilution multiple)	Dye industry	50	180	–
		Other discharge units	50	80	–
3	Suspended solids (SS)	Mining, mineral processing, coal preparation industry	100	300	–
		Vein gold concentration	100	500	–
		Gulch-gold concentration in remote areas	100	800	–
		Secondary sewage treatment plant in the town	20	30	–
		Other discharge units	70	200	400
4	Biochemical oxygen demand after five days (BOD_5)	Sugarcane sugar making, ramie degumming, wet fiberboard industry	30	100	600
		Beet sugar making, alcohol, monosodium glutamate, leather, chemical fiber pulp industry	30	150	600
		Secondary sewage treatment plant in the town	20	30	–
		Other discharge units	30	60	300
5	Chemical oxygen demand (COD)	Beet sugar making, coking, fatty acid synthesis, wet fiberboard, dye, wool washing, the organophosphorus pesticide industry	100	200	1000

(continued)

Table 12 (continued)

Number	Pollutant	Application	First-class standard	Second-class standard	Third-class standard
		Monosodium glutamate, alcohol, pharmaceutical raw materials, biopharmaceutical, ramie degumming, leather, chemical fiber pulp industry	100	300	1000
		Petrochemical Industry (including petroleum refining)	100	150	500
		Secondary sewage treatment plant in the town	60	120	–
		Other discharge units	100	150	500
6	Petroleum	All discharge units	10	10	30
7	Animal & vegetable oils	All discharge units	20	20	100
8	Volatile phenol	All discharge units	0.5	0.5	2.0
9	Total cyanide	Film development (iron cyanides)	0.5	5.0	5.0
		Other discharge units	0.5	0.5	1.0
10	Sulfide	All discharge units	1.0	1.0	2.0
11	Ammonia nitrogen	Pharmaceutical raw materials, dyes, petrochemical industry	15	50	–
		Other discharge units	15	25	–
12	Fluoride	Yellow phosphorus industry	10	20	20
		Low fluoride area (water fluoride content < 0.5 mg / L)	10	20	30
		Other discharge units	10	10	20
13	Phosphate (as phosphorus)	All discharge units	0.5	1.0	–
14	Formaldehyde	All discharge units	1.0	2.0	5.0

(continued)

Table 12 (continued)

Number	Pollutant	Application	First-class standard	Second-class standard	Third-class standard
15	Anilines	All discharge units	1.0	2.0	5.0
16	Nitrobenzene	All discharge units	2.0	3.0	5.0
17	Anionic surfactant	Synthetic detergent industry	5.0	15	20
		Other discharge units	5.0	10	20
18	Total copper	All discharge units	0.5	1.0	2.0
19	Total zinc	All discharge units	2.0	5.0	5.0
20	Total manganese	Synthetic fatty acid industry	2.0	5.0	5.0
		Other discharge units	2.0	2.0	5.0
21	Color developer	Film development	2.0	3.0	5.0
22	Total amount of developer and oxide	Film development	3.0	6.0	6.0
23	Phosphorus	All discharge units	0.1	0.3	0.3
24	Organophosphorus pesticides (as phosphorus)	All discharge units	N.D	0.5	0.5
25	Fecal coliform count	Pathogenic wastewater from hospitals*, veterinary hospitals, and medical institutions	500 colonies/L	1000 colonies/L	5000 colonies/L
		Sewage from infectious disease and tuberculosis hospital	100 colonies/L	500 colonies/L	1000 colonies/L
26	Total residual chlorine (chlorinated hospital sewage)	Pathogenic wastewater from hospitals*, veterinary hospitals, and medical institutions	< 0.5**	(contact time ≥ 1 h)	(contact time ≥ 1 h)
		Sewage from infectious disease and tuberculosis hospital	< 0.5**	(contact time ≥ 1.5 h)	(contact time ≥ 1.5 h)

* Hospitals with more than 50 beds

Note ** After chlorination and disinfection, denitrification shall be performed to meet this standard

Article 4.1.2 Sewage discharged into Class IV, and V waters in the *Environmental Quality Standards for Surface Water* (GB 3838–2002) and Class III waters in the *Sea Water Quality Standard* (GB 3097–1997) shall meet the second-level standards.
Article 4.1.3 Sewage discharged into the urban drainage system with secondary sewage treatment plants shall meet the third-level standards.
Article 4.1.4 Sewage discharged into the urban drainage system without secondary sewage treatment plant must comply with the provisions of 4.1.1 and 4.1.2 according to the receiving water area's functional requirements.
Article 4.1.5 In the protected areas designated in Class I and II waters in the *Environmental Quality Standards for Surface Water* (GB 3838–2002) and Class I sea areas designated in the *Seawater Quality Standard* (GB 3097–1997), it is forbidden to build new sewage outlets. The total amount of pollutants in the existing outlets should be controlled according to the functional requirements to ensure that the receiving water quality meets the water quality standards for specified purposes.

3. Pollutant discharge standards for sewage treatment plants
Discharge standard of pollutants for municipal wastewater treatment plant (GB 18,918–2002) stipulates that according to the environmental function and protection target of surface water discharged from municipal sewage treatment plant and the treatment process, the general pollutant standard of the basic control project is divided into first-level standard, second-level standard, and third-level standard. The first-level standard is divided into A standard and B standard.

(1) The A standard is the basic requirement of the effluent from the municipal sewage treatment plant as the reuse water. When the effluent is introduced into rivers and lakes with low dilution capacity for urban landscape water and general reuse water, the first-level A standard should be implemented.
(2) When effluent from urban sewage treatment plant is discharged into Class II functional waters of surface water (excluding designated protection areas for drinking water sources and swimming areas) designated in the *Environmental Quality Standards for Surface Water* (GB 3838–2002) or Class II functional waters of seawater designated in the *Seawater Quality Standard* (GB 3097–1997) and enclosed or semi-enclosed waters such as lakes and reservoirs, the first-level B standard should be implemented.
(3) When effluent from the urban sewage treatment plant is discharged into the classes IV and V functional waters of surface water designated in the *Environmental Quality Standards for Surface Water* (GB 3838–2002) or the class III and class IV functional sea areas designated in the *Seawater Quality Standard* (GB 3097–1997), the second-level standard shall be implemented.
(4) According to the local economic conditions and water pollution control requirements, the sewage treatment plants in the constructed towns with

non-key control basins and non-conservation areas shall implement the third-level standard when adopting the first-level enhanced treatment process. However, sewage treatment plants must reserve the room for second-level treatment facilities to meet the second-level standard by stages.

The maximum allowable discharge concentration (daily average value) of the basic control project from urban sewage treatment plants is in Table 13.

Onsite evaluation or supporting materials:

Testing reports and account.

Table 13 Maximum allowable discharge concentration (daily average value) of the basic control project Unit: mg/L

Number	Basic control project		First-level standard		Second-level standard		Third-level standard
			Standard A	Standard B			
1	Chemical oxygen demand (COD)		50	60	100		120*
2	Biochemical oxygen demand after five days (BOD_5)		10	20	30		60*
3	Suspended solids(SS)		10	20	30		50
4	Animal and vegetable oils			1	3	5	20
5	Petroleum			1	3	5	15
6	Anionic surfactant			0.5	1	2	5
7	Total nitrogen (in nitrogen)			15	20	–	–
8	Ammonia nitrogen (as nitrogen) **			5 (8)	8 (15)	25 (30)	–
9	Total phosphorus (as phosphorus)	Completed before December 31, 2005		1	1.5	3	5
		Constructed after January 1, 2006		0.5	1	3	5
10	Chroma (Dilution multiple)			30	30	40	50
11	pH			6 ~ 9			
12	Fecal coliform count (colonies/L)			10^3	10^4	10^4	–

Note * The removal rate shall be more than 60% when the influent cod is more than 350 mg / L, and the removal rate shall be more than 50% when BOD is more than 160 mg / L

** The value outside the bracket is the control index when the water temperature is higher than 12 °C, and the value in the bracket is the control index when the water temperature is ≤2 °C.

57 Industrial wastewater discharge

Standard score: 15

Instructions: The government encourages zero discharge of industrial wastewater. If the discharge is necessary, 15 points are rewarded for water quality reaching the standard

Assessment methods: Search information and check the site

Criteria and standard: Inspection data from the environmental protection department

See Article 56 for reference.

Onsite evaluation or supporting materials:

Test reports and accounts.

58 Drainage pipeline settings

Standard score: 10

Instructions: Ten points are rewarded for the cleaning pipeline laid, rainwater, and sewage pipe groups set separately

Assessment methods: Check the site

Explanations: This article is the requirement for cleaning and diverting in drainage management.

This item complies with the *Regulations on Urban Drainage and Sewage Treatment.*

Article 19 of the *Regulations on Urban Drainage and Sewage Treatment* stipulates that except for arid areas, rainwater and sewage diversion shall be implemented in new areas construction; and for areas where rainwater and sewage are combined, rainwater and sewage diversion shall be carried out in accordance with the requirements of urban drainage and sewage planning. The flow diversion of rainwater and sewage can be conducted simultaneously with the old urban area reconstruction and road construction.

In rainwater and sewage diversion areas, rainwater and sewage piping networks should not be mixed to construct new areas and reconstruct old urban areas.

In areas where conditions permit, it is necessary to promote initial rainwater collection and treatment to determine the interception multiple. The discharge control and pollution prevention of initial rainwater can be strengthened by constructing initial rainwater storage pools and intercepting main pipes.

1. Clean water, rainwater and sewage separation

   The system of mine wastewater collection, treatment, and utilization is perfect with complete functions; the sewage pipeline is laid separately, and the rainwater and sewage pipe groups are set separately; the surface runoff water collection and utilization system is built in the open pit.

Rainwater and sewage diversion is a sewage discharge method separating rainwater from domestic sewage and conveying them in one pipeline for discharge or subsequent treatment. Rainwater and sewage diversion effectively reduce sewage amounts and is convenient for rainwater collection and utilization, centralized management, and discharge.

Clean-water and sewage separation separates highly polluted water from non-polluted or low-polluted water to reduce the pollutants amount and the cost of water treatment. Rainwater and sewage double ditches can be built on the main roads along the mountains to separate the rainwater from road sewage.

The production drainage, rainwater, and domestic sewage in mining areas should realize clean water, rainwater, and sewage separation.

2. Relevant requirements

   (1) Clean water, rainwater, and sewage separation can save the cost of sewage treatment. The sewage pipeline should be laid separately, and the rainwater ditches and sewage pipe groups should be set separately.
   (2) Rainfall sewage is discharged through pipelines and diverted on rainy days. Rainwater flows into rivers or gets reused through road sewers. Sewage is transported to sewage treatment plants.
   (3) Clean water and sewage separation is beneficial to the comprehensive utilization of mine water. The workplace's discharge water is heavily polluted, but the water volume is small and easy for centralized treatment. A downhole sedimentation tank can be built to discharge the water separately. Clean water inrush from the well can be directly applied, and acid mine sewage should be treated separately according to water quality characteristics.

Onsite evaluation or supporting materials:

Check the design scheme of the mine drainage system.

59 Discharge requirements for surface runoff and leaching water

Standard score: 15

Instructions: (1) Rainwater interception (drainage) ditch is built in the mine area, and sedimentation tank and water intake equipment are constructed. The collected surface runoff water, leaching water are discharged or reused after being settled. Ten points are rewarded for meeting these requirements. (2) Five points for setting interception (drainage) ditches in the waste rock dump

Assessment methods: Check the site

Criteria and standard: The mining area's overall design

Explanations: This article is the requirement for surface runoff and leaching water discharge.

Rainwater interception (drainage) ditch, sedimentation tank, and water intake equipment are built in the mine area. The collected surface runoff water, leaching water are discharged or reused after being settled. In the low-lying area of the production area, the initial rainwater collection tank is built for rainwater precipitation and clarification, after which the initial rainwater can be used for industrial sites and roads' dust suppression and watering.

Mining areas and storage yards (ore stacking yards) shall construct rainwater interception (drainage) ditches and sedimentation pools in accordance with *Technical Specifications for Comprehensive Control of Soil and Water Conservation - Small Engineering of Store, Drainage, and Draw Water* (GB/T 16,453.4–2008). The collected surface leaching water shall meet the corresponding water quality requirements of *Environmental Quality Standards for Surface Water* (GB 3838–2002).

1. Summary

   (1) Drains

   Drains refer to the drainage ditch (drainage ditch at the edge of subgrade, which is mainly used to collect and drain a small amount of surface water within the scope of subgrade and flowing to the subgrade), intercepting ditch, and the water ditch leading water outside the subgrade, bridge and culvert, farmland and residential land. It is generally set under the filling subgrade to drain the pavement water.

   The natural drainage ditch is excavated by the construction site's natural conditions, saving a lot of time and labor. Laying pebbles or gravel at the bottom of the ditch can not only avoid the scour of the current, help protect the bottom of the ditch but can deposit the dirt in the water as well. The natural drainage ditch is often adopted in construction projects with a small scale, short construction period, or when water or rainwater is not large during construction.

   The open drainage ditch is usually made combining with the hardened ground on site. It has advantages of labor, material, and economic saving. The function of the open drainage ditch is also apparent. The disadvantage is that the section should not be extensive, especially when the construction site is small, the drainage is difficult due to the limitation of the depth and size. Simultaneously, the open drainage ditch is easier to affect or hinder construction personnel and machinery's motions. However, for those construction projects with short construction periods, small construction scales, the open drainage ditch has specific suitability.

   (2) Intercepting ditches

   The intercepting ditch is generally set on the subgrade slope in the excavation section to intercept the subgrade's surface water erosion. It mainly intercepts the water flowing to the subgrade on the hillside and is set outside the top of the cutting slope.

The intercepting ditch at the top of the excavated subgrade shall be set at 5 m away from the groove, and the distance between the intercepting ditch on the upper side of the filled subgrade and the filling slope toe shall not be less than 2 m. In rainy areas, one or more intercepting ditches can be set up according to the actual situation. Its function is to intercept the surface water flowing from the upper position to the subgrade, protect the excavation slope, and fill the slope's foot from being scoured by the current.

The location of the intercepting ditch is the main factor. In the case of no dump, the distance between the edge of the intercepting ditch and the top of the excavated subgrade slope depends on the soil quality, with the principle of not affecting the slope stability. When there is a dump above the subgrade, the intercepting ditch should be 1–5 m away from the dump; the toe of the dump should not be less than 10 m away from the top of the excavation slope; a cross slope with 2% degrees inclining to the intercepting ditch should be set at the top of the dump. The embankment's intercepting ditch on the hillside shall be at least two meters away from the toe of the embankment, and the soil excavated from the intercepting ditch shall be filled between the embankment and the intercepting ditch. The revetment or soil platform with a slope of 2% toward the ditch shall be built to make the surface water flow into the intercepting ditch and drain out.

2. Layout principles of intercepting ditches

(1) When the slope's length without measures is too long, several intercepting ditches should be added to the slope. The distance of other intercepting ditches is generally 20–30 m, determined by design calculation according to the ground slope, soil quality, and rainstorm runoff.

(2) The water-retaining interception ditch is arranged along the contour line. The water-draining interception ditch should take a 1–2% gradient from the contour line (also known as slope and gradient, which refers to the ratio of the elevation difference between any two endpoints horizontal distance between the two points).

(3) When the intercepting ditch is not horizontal, a 20–30 cm high small soil retaining wall should be built every 5–10 m to prevent scouring.

(4) The drainage end of the water-draining interception ditch shall be connected with the slope drainage ditch. Scour prevention measures shall be taken outside the connection.

3. Layout principles of drains

(1) Drainage ditches are generally arranged at the two ends or lower ends of slope interception ditches to exclude surface runoff that cannot be

accommodated by the interception ditches. The end of a drainage ditch connects to a reservoir or a natural drain.

(2) The slope gradient of drainage ditches depends on the drainage outlet location (reservoir or natural drainage channel). When the drainage outlet is at the slope's foot, drainage ditches are arranged roughly orthogonal to the slope's contour; when the drainage outlet is on the slope, drainage ditches can be arranged basically along the contour or obliquely intersected with the contour. All layouts must be scour-proof (turf or stone lining).

4. Layout principles of reservoir and sand basin

(1) Reservoirs are generally located at the foot of the slope or part of the low pit on the slope surface and connected with the terminal of drainage ditch (or water-draining interception ditch) to store drainage on the slope surface.

(2) The reservoir's distribution and capacity are determined by a total runoff on the slope, the relationship between storage and discharge, the construction convenience, and the local conditions. The engineering system of storage and discharge on a slope can either centralize a water tank or distribute several tanks. A single cell's capacity varies from hundreds of cubic meters to tens of thousands of cubic meters.

(3) The reservoir's location should be determined by favorable terrain, good lithology (no cracks, caves, and gravel layer), large water storage capacity, a small quantity of work, convenient construction, and other conditions.

(4) The sand basin is generally located near the upstream of the water tank's intake. Water discharged from drainage ditches (or water-draining interception ditch) goes into the sand basin first, and then fresh water is discharged into the basin after the sand has settled.

(5) The sand basin's location can be determined according to local topography and engineering conditions, either close to the reservoir or at a distance from the reservoir.

5. Layout of intercepting (drainage) ditches in mining areas

The interception layout (drainage) ditch in a mining area shall be designed according to the mining area's general layout and the water and soil conservation scheme. In assessing green mines, it is necessary to consider whether the interception (drainage) ditch at the waste dump site and the interception (drainage) ditch on both sides of the road are suitable. At the same time, the waste stacking site should be considered:

(1) Slag retaining wall should be set at the downstream of waste rock yard and waste slag yard.

(2) In open pits, waste rock yards, waste slag yards, upstream and sides of tailings reservoirs, suitable flood interception and drainage facilities shall be provided according to actual topographical conditions.

Onsite evaluation or supporting materials:

(1) Rainwater interception (drainage) ditch design.
(2) Check on-site rainwater interception and drainage as well as the construction of reservoir and sand basin.

4 Solid Waste Discharge

60 Solid waste discharge requirements

Standard score: 30

Instructions: For the solid waste that cannot be comprehensively utilized: (1) Ten points are rewarded for dividing the hazardous waste, general waste, and domestic waste into different categories; (2) Twenty points for disposing the solid waste according to the national standards, or entrusting a third party qualified unit to dispose of the solid waste

Assessment methods: Search information and check the site

Criteria and standard: *Standards for Pollution Control on the Storage and Disposal Site for General Industrial Solid Wastes* (GB18599), *Standard for Pollution Control on Hazardous Waste Storage* (GB18484、18,597、18,598), *Standard for Pollution Control on the Landfill Site of Municipal Solid Waste* (GB18485 and 16,889)

Explanations: This article is the requirement for solid waste discharge.

1. Solid-waste classified treatment

   (1) Solid wastes from office buildings, living places, laboratories are uniformly put into the garbage bins and transported to the garbage station by environmental protection vehicles for treatment. The result accords with the national and local standards.
   (2) Solid wastes of each working face on the construction site are put into the garbage bin for classification and centralized collection and finally transported to the garbage station or for unified treatment.
   (3) Solid wastes such as steel slag, boiler slag, fly ash, gangue, industrial dust, waste residue, sludge, domestic waste, and hazardous waste shall be classified and managed.

2. Solid-waste pollution

   Untreated factory wastes and domestic wastes are simply piled up in the open air, occupying land and damaging the landscape. Moreover, the waste's harmful components are transmitted in the air through the wind and enter the soil, river, or underground water source through the rain. This process is solid-waste pollution.

(1) Contaminate water. Without harmless treatments, solid wastes will enter rivers and lakes along with natural precipitation or surface runoff, resulting in long-term sedimentation and water surface reduction, deteriorating the contamination. Harmful components of solid waste, such as mercury (from red plastic, neon lamp tube, battery, and vermilion inkpad), cadmium (from printing, ink, fiber, enamel, glass, cadmium pigment, paint, and colored ceramics), lead (from yellow polyethylene, lead water pipe, and antirust coating) and other trace toxic elements, if not handled properly, can enter the soil with the leaching water and pollute the groundwater. Simultaneously, these harmful components will infiltrate into the water network with rainwater, flow into wells, rivers, and even the nearby sea areas. Absorbed by plants, they will enter the human body through the food chain, affecting human health. Pollution indexes such as concentration, color, total bacterial count, and heavy metal content of groundwater around garbage landfills in some cities in China exceed the standard exceptionally.

(2) Contaminate the air. The dry matter or lightweight of solid waste will float in the wind, causing air pollution. Incineration is a popular way to treat solid waste at present, but it will produce many harmful gases and dust. Microorganisms will decompose some organic solid wastes under suitable temperature and humidity for a long time, and harmful gases will be released simultaneously.

(3) Contaminate the soil. Soil is the place where many bacteria, fungi, and other microorganisms live together. These microorganisms play an essential role in the manifestation of soil functions. They constitute a balanced ecosystem with the soil. However, untreated harmful solid wastes, through weathering, rain, surface runoff, and other effects, their toxic liquid will infiltrate into the soil, kill the soil's microorganisms, and destroy the soil's ecological balance. Seriously polluted places will be barren.

(4) Occupy the land. Solid wastes are growing quite rapidly so that many cities use large areas of farmland to pile them up. The satellites' photos show that the vast white garbage surrounding the city is so conspicuous.

3. Solid-waste treatment

Solid-waste treatment usually refers to converting solid wastes into materials suitable for transportation, storage, utilization, and disposal by physical, chemical, biological, physical, and biochemical methods. Solid-waste treatment aims at harmlessness, reduction, and resource utilization. At present, the main methods include compaction, crushing, sorting, solidification, incineration, biological treatment, and others.

(1) Compaction. Compaction is a pretreatment technology to reduce transportation costs and prolong the landfill life by reducing the volume of waste. Solid wastes such as automobiles, cans, and plastic bottles are usually compacted to reduce volume, while some wastes that may cause

operational problems such as tar, sludge, or liquid materials, are not suitable for compaction.

(2) Crushing. For reducing the shape of solid waste entering incinerator, landfill, and composting system, it is necessary to crush the solid waste in advance. Wastes treated by crushing have uniform size and uniform texture due to the elimination of large voids. There are many crushing methods for solid wastes, including impact crushing, shear crushing, extrusion crushing, and friction crushing. Besides, there are special low-temperature crushing and mixed crushing.

(3) Sorting. Solid waste sorting is an important means to realize solid-waste recycling and reduction. One way is to separate the useful wastes from the harmful ones by sorting; the other way is to separate the waste with different particle sizes. The basic principle of separation is to use the difference in some aspects of materials. For example, the magnetic and non-magnetic differences, particle size differences, and density differences in waste materials are used for separation.

(4) Solidification. Solidification technology is a harmless treatment process by adding solidified base material to waste so that hazardous solid waste is fixed or contained in inert solidified base material. The solidification products should have good anti permeability, good mechanical properties, anti-leaching, anti-drying, and anti-freezing properties. According to the different base materials, solidification can be divided into sinking solidification, asphalt solidification, glass solidification, and resin solidification.

(5) Incineration and pyrolysis technology. Incineration is a comprehensive treatment process with high-temperature decomposition and deep oxidation of solid wastes. The advantage of incineration is that a large number of hazardous wastes are decomposed into harmless substances. However, the incineration method also has some disadvantages, including large investment, secondary pollution caused by smoke exhaust in the incineration process, and severe equipment corrosion. Pyrolysis decomposes organic matter into gas, liquid, and solid products by heating at high temperatures (500–1,000 °C) with no oxygen or oxygen deficiency. Compared with incineration, pyrolysis is a more promising treatment method, and its most significant advantage is less capital investment.

(6) Biological treatment. Biological treatment technology uses microorganisms to decompose organic solid waste to make it harmless. It can transform organic solid wastes into energy, food, feed, and fertilizers and extract metal from wastes and waste residue. The technology is a significant technical method for solid-waste recycling. Composting, biogas, saccharification of waste cellulose, feed conversion of waste fiber, and bioleaching are widely used.

4. Hazardous waste management
   Sign a recycling and disposal contract with a qualified recycling unit, and regularly recycle and dispose of solid wastes such as waste oil barrels, waste paint barrels, and other hazardous wastes. The temporary storage room shall be provided with various signs, anti-seepage, and ventilation devices according to the environmental protection requirements.

Onsite evaluation or supporting materials:

Solid-waste reduction measures.

5 Noise Emission

61 List of main noise points
Standard score: Five
Grading description: The mine has a list of sites generating the main noise (sound) and their posts. If necessary, an onsite test can be carried out, and five points will be rewarded for meeting the requirements
Assessment method: Check the site

[Interpretation] This article requires mining enterprises to organize a list of noise places.

Noise is environmental pollution, considered to be the third-largest public hazard after air pollution and water pollution. Noise pollution refers to the phenomenon that the environmental noise generated exceeds the state's environmental noise emission standard and interferes with others' normal work, study, and life. Although the intensity of noise in daily life may not cause death to humans or animals, it can endanger people's health. Countries worldwide pay much attention to noises and list noise pollution as one of the main environmental hazards.

1. Classification of mine noise sources
   (1) According to the location of noise, it can be divided into surface noise sources and underground noise sources.
   (2) According to the causes of noise, it can be divided into equipment noise sources and non-equipment noise sources: ① Equipment noise sources include fans, air compressors (referred to as AC), rock drilling equipment, loading and unloading equipment, transportation equipment, and crushing equipment. ② Non-equipment noise sources include blasting, discharge, and leakage of compressed gas in compressed gas pipelines, rib spalling, roof fall and cover caving, as well as rolling and impact noise of ore dumped into silos and chutes.

2. High Noise Equipment
   The noise mainly includes the noise generated by mechanical equipment during ore mining, crushing, shipment, ore processing, and vehicles' traffic noise and blasting noise generated in the blasting process.
   High noise equipment mainly includes rock drilling, crushing equipment, screening equipment, fans, air compressors, drainage pumps, and transportation equipment.
   Mining enterprises should make a list of workplaces where high-noise equipment is located and gradually improve them.
3. Noise Requirements on Post
   Article 5 of the *Industrial Enterprise Noise Hygiene Standard (Trial Draft)* requires that the noise standard in production workshops and industrial enterprises' workplaces should be 85 dB. Existing industrial enterprises can temporarily relax the standards if they fail to temporarily meet the standards but must not exceed 90 dB. *Occupational Exposure Limits for Harmful Factors in the Workplace Part 2: Physical Factors* (GBZ 2.2—2007) specifies the noise exposure limits in the workplace as follows: The exposure limit for 8 h per day is 85 dB; the exposure time is 4 h per day; the exposure limit for 4 h is 88 dB per day; the exposure limit for 2 h is 91 dB per day; the exposure limit for 1 h is 94 dB per day;
   Article 10 of the *Industrial Enterprise Noise Hygiene Standard (Trial Draft)* requires that in existing industrial enterprises, where the noise exceeds the requirements of this standard in production workshops and workplaces, effective control measures must be taken to meet the requirements of this standard within a time limit. Before meeting the standards, factories and mining enterprises must issue personal protective equipment to protect workers' health.
4. List of noise points
   Table 14 shows the list of noise points.

Onsite evaluation or supporting materials:

List of main noise points.

Table 14 Examples of noise point lists

Workplace	Major equipment	Contact time on post/h	Exposure limit/ Db(A)
Drilling post and surrounding workplace	Down-hole drill	8	85
Excavator	Loader	8	85
Crushing workshop	Crusher and vibrating screen	8	85

62 Noise disposal requirements

Standard Score: 15

Score Description: Perform noise reduction treatment on high-noise equipment such as rock drilling, crushing, and air pressure. Fifteen points are awarded for noise suppression, vibration reduction, and vibration isolation

Assessment method: Check the relevant monitoring report

[Interpretation] This article is a requirement for noise disposal.

It mainly adopts sound absorption, sound insulation, vibration reduction, vibration isolation, and other technical measures to reduce the noise of high-noise equipment. The specific methods are as follows.

(1) High-noise equipment should be arranged in a relatively concentrated area away from the management area and living area.
(2) High-noise workshops should be arranged separately from low-noise workshops.
(3) The production facilities that produce high noise should be arranged around buildings and structures that are relatively insensitive to noise, tall, and oriented in favor of sound insulation.
(4) Adopt the sinking design according to the actual situation and local conditions, and add shock absorption and muffler devices for noise reduction.
(5) Partially or integrally encapsulate high-noise equipment in the key parts with apparent noise or near the factory boundary, and install sound-absorbing materials in sound insulation screens and sound insulation enclosures.
(6) The common method to control fan noise is to install resistive mufflers at the fan's inlet and outlet. Use a sound insulation box for occasions with higher noise reduction requirements, and install a shock absorber between the unit and the foundation.
(7) The measures to reduce the air compressor's noise are to design and install sound insulation covers according to local conditions and install mufflers at the air inlet and outlet.
(8) The measures to control the motor noise include installing a sound insulation cover, changing the motor cooling fan structure, such as changing the diameter of leaf blades and wind blades, reducing the number of leaf blades, and installing mufflers.
(9) The measures to control the noise of rock drilling equipment include installing a light muffler at the rock drill's exhaust outlet, making drill set with foam neoprene, sealing the shell of the rock drill body with a noise insulation cover, and setting sound insulation operation room on the trolley.
(10) The measures to control the noise of the main drainage pump include installing sound-absorbing and soundproofing panels around the walls of the pump

room and on the top of the machine room, and fixing them with light steel keels. The holes connecting the machine room and other rooms are tightly blocked to avoid air transmission; ensure the airtight of door and windows in the machine room; add a soundproof cover to the water pump; install the vibration isolation platform at the bottom of the pump.

Onsite evaluation or supporting materials:

Review relevant designs or measures.

63 Noise emission requirements

Standard score: Ten

Score description: Ten points are rewarded for reaching the noise emission standard at the factory boundary

Assessment method: Search information

According to the file: *Environmental Noise Emission Standard at the Boundary of Industrial Enterprises* (GB12348—2008)

[Interpretation] This article is a requirement for noise emission.

1. Emission requirements

   The noise in the mining area is in accordance with the *Environmental Noise Emission Standard at the Boundary of Industrial Enterprises* (GB 12,348—2008) for Class 2 acoustic environment function zone standards, namely 60 dB at daytime and 50 dB at night; the standard limit from *Railway Noise Limits and Measurement Methods* (GB 12,525) —1990), namely 70 dB during the day and 70 dB at night.

   (1) Environmental noises at the boundary of industrial enterprises shall not exceed the emission limits (See Table 15).
   (2) The maximum sound level of frequent noise at night shall not exceed the limit value of 10 dB (A).

Table 15 Limits of environmental noise emission at the boundary of industrial enterprises Unit: dB (A)

Acoustic environment function category outside the factory boundary	Day	Night
0	50	40
1	55	45
2	60	50
3	65	55
4	70	55

(3) The maximum sound level of occasional noise at night shall not exceed the limit value of 15 dB (A).

(4) If an industrial enterprise located in areas without acoustic environmental function zones, when there are noise-sensitive buildings outside the factory boundary, the local people's government at or above the county level shall determine the acoustic environment quality requirements for areas outside the plant boundary referring to the *Acoustic Environment Quality Standard* (GB 3096—2008) and *Technical Specifications for the Division of Acoustic Environment Functional Areas* (GB/T 15,190–2014), and implement the corresponding environmental noise emission limits at the factory boundary.

(5) When the distance between the factory boundary and the noise-sensitive building is less than 1 m, the environmental noise of the factory boundary shall be measured indoors in the noise-sensitive building, and the corresponding limit in Table 16 shall be reduced by 10 dB(A) as the evaluation basis.

The acoustic environment functional area's application scope is divided into the following five types according to the area's functional characteristics and environmental quality requirements.

Type-0 acoustic environment functional areas require quietness, such as rehabilitation areas.

Type-1 acoustic environment functional areas need to be quiet for residential, medical and health care, cultural education, scientific research and design, and administrative office as their main functions.

Type-2 acoustic environment functional areas refer to areas where commercial finance and market trade are the main functions, or where residence, commerce, and industry are mixed, and the residence needs to be quiet.

Type-3 acoustic environment functional areas take industrial production, warehousing, and logistics as the main function, and industrial noise needs to be prevented from having a severe impact on the surrounding environment.

Type-4 acoustic environment functional areas are within a certain distance on both sides of the traffic trunk line that needs to prevent traffic noise from causing severe impact on the surrounding environment, including two types: 4a and 4b. Type 4a includes expressways, primary highways, secondary highways, urban expressways, urban trunk roads, urban secondary trunk roads, urban rail transit (ground sections), and areas on both sides of the inland waterway; type 4b includes areas on both sides of a trunk railway.

2. Examination requirement

Noise at the industrial site boundary shall meet the corresponding noise limit at the boundary according to *Environmental Noise Emission Standard at the Boundary of Industrial Enterprises* (GB 12,348–2008). Underground mines are detected at industrial sites' boundaries, and open-pit mines are detected at mining areas' boundaries.

Onsite evaluation or supporting materials:

Relevant monitoring report.

Chapter 8
Scientific and Technological Innovation and Intelligent Mines

Scientific and technological innovation is an important driving force for the construction of green mines in the green mine construction process, which technically guarantees cost reduction and efficiency improvement; intelligent mines reflect how to make production at the optimal level, how to maintain the coordinated relationship between resource development and ecological protection.

This chapter's assessment focuses on considering the enterprise's technological innovation ability and intelligent mine's construction level.

1 Scientific and Technological Innovation

64 Technology R&D team

Standard score: Three

Score description: The enterprise builds the technical research and development troop, and three points are rewarded for full-time technicians

Assessment method: Search information

Basis or standard: Science and technology management system

[Interpretation] This article requires mining enterprises to build technical research and development teams.

The basic tasks of scientific and technological innovation of mining enterprises include establishing scientific and technological R&D teams, and promoting scientific and technological achievements, technological transformation, and green development of mining enterprises and industrial transformation and upgrading. These four requirements also reflect the position of scientific and technological innovation

J. Deng et al., *Interpretation of Green Mine Evaluation Index*,
https://doi.org/10.1007/978-981-16-5433-6_8

in the construction process of green mines and clarify technological innovation's role in constructing green mines.

Establishing a scientific and technological R&D team and equipping full-time management personnel is the basis for scientific and technological innovation.

Onsite evaluation or supporting materials:

Check the organizational setting of the enterprise, and the enterprises should provide relevant documents.

65 Technology research and development management system

Standard score: Three

Score description: Three points are awarded for technical research and development and management system

Assessment method: Search information

Basis or standard: Science and technology management system

[Interpretation] This article requires mining enterprises to establish a scientific and technological management system.

Mining enterprises must establish a scientific and technological innovation system, study key technologies supporting their green development, and set up scientific and technological management systems to improve the technology.

The science and technology management system mainly includes the sources and procedures of scientific and technological funds, scientific research project management, appraisal of scientific and technological achievements, and scientific and technological achievements management.

Onsite evaluation or supporting materials:

Review the technology management system of enterprises.

66 Collaborative innovation system

Standard score: Six

Score description: Establish industry-university-research-application collaborative innovation system

① Establish technological innovation cooperation relationships with scientific research institutes and universities. Two points will be rewarded for signing cooperation agreements to establish enterprise technology platforms, including engineering technology centers, enterprise technology centers, key laboratories, academician expert workstations, and innovation studios

② Carry out technical research supporting the development of the enterprise's main business, and two points are rewarded for project establishment documents or project ledger materials

③ Two points are rewarded for improving the technological level of the enterprise and supporting materials

Assessment method: Search information

Basis or standard: Announcement documents of the competent authority, project approval documents, and project ledger

[Interpretation] This article requires establishing scientific and technological innovation systems combining production, learning, research, and application.

"Production, learning, research, and application" refers to the cooperation of enterprises, universities, research institutes, and users to play their respective advantages to form a powerfully advanced system integrating research, development, production, and application. It is a high degree of synergy and integration of functions and resource advantages in different social divisions of labor in scientific research, education, production, and markets, a seamless connection between technology and industry, and cooperative system engineering. It can solve the disconnection between scientific and technological innovation and economic and social development, the lack of compatibility between universities' discipline directions and leading local industries, and the distance between scientific research teams and the market. Combine the methodology mastered by universities, the phased leading technology mastered by scientific research institutes, field personnel's ability to find problems, and the role of time in solving problems. Make full use of the advantages of talents at all levels and gain practical experience to combine production and scientific research based on practical ability organically. "Four Forces", power in enterprises, magic in the market, ability in an institute, and onsite energy, are integrated into one unit with vital creativity.

Establishing a technological innovation system combining "production, learning, research, and application" is the only way to improve independent innovation capabilities. It can accelerate the transformation and upgrading of the industrial structure and realize the revolutionary transfer and upgrading of China's comparative advantages, that is, from relying solely on low-cost and cheap labor to the innovative power formed by the integration and simplification of enormous "industry-university-research" resources. Scientific and technological innovation can be divided into scientific and technological research and development, technological transformation, and training exchange.

(1) Science and technology research and development category: it is the general term for original scientific research and technological innovation, and a process of creating and applying new knowledge, new technology, and new process, adopting new production mode and management mode to develop new products, improve product quality and provide new services. Scientific research and development innovation can be divided into three aspects: knowledge innovation, technological innovation, and management innovation led by modern technology.

(2) Technical modification: The transformation of existing facilities, production process conditions, and auxiliary facilities by adopting domestic and foreign advanced and new technologies, new equipment, new processes, and new materials is called technological transformation. Technological transformation refers to applying scientific and technological achievements in all aspects of enterprise products on the premise of adhering to scientific and technological progress. Using advanced technology and equipment to replace backward technology and equipment can expand production and achieve production safety, a protected environment, increased variety, improved quality, saving energy, reduced consumption of raw materials, increased labor productivity, and increased economic benefits.

(3) Training and communication: The mine enterprises invite relevant industry experts to carry out technical training for mining professionals according to their conditions; carry out field inspections to advanced peer units, learns advanced technological achievements, and exchange experience in scientific and technological innovation; cooperate with professional technical teams in the form of technical collaboration to enhance the strength of technological innovation.

The "production, study, research and application" cooperation between mining enterprises and universities and research institutes also focuses on the above three considerations. Strengthen the cooperation with universities, scientific research institutions, and technology-based organizations, sign the "production, learning, research and application" cooperation agreement, and build the "production, learning, research and application" collaborative innovation system.

Onsite evaluation or supporting materials:

(1) Cooperation agreement of "production, learning, research, and application"
(2) Enterprise technology platform, including engineering technology center, enterprise technology center, key laboratory, and academician workstation.

67 Science and technology awards

Standard score: 18

Grading description: Eighteen points are rewarded for enterprises' research projects or achievements for national awards, 12 points for provincial and ministerial awards, and ten points for the *Social Science and Technology Awards Catalogue* of the National Awards Office. All awards should promote green mine construction and reflect the unit's name, with a total score of no more than 18 points

Assessment method: Search information

Basis or standard: Announcement documents of a competent authority, project approval documents, and project ledger

[Interpretation] This article encourages mining enterprises to declare their scientific and technological innovation achievements for relevant awards to be recognized by society or the industry.

1. National awards
   The National Natural Science Award, the National Technological Invention Award, and the National Science and Technology Progress Award shall be issued by the State Council with certificates and bonuses. The People's Republic of China International Science and Technology Cooperation Award shall be issued by the State Council.
2. Provincial-level awards
   Provincial and ministerial awards here refer to awards such as science and technology awards, invention awards, and natural science awards directly awarded by the party committees or people's governments of the provinces, autonomous regions, and municipalities directly under the Central Government of the People's Republic of China, or directly awarded by the Ministry of Education, the Ministry of Culture, the Ministry of Public Security, and the National Defense Technology Industry Directly awarded by the Bureau and other national ministries and commissions and the People's Liberation Army.
3. Other awards
   The awards in the *Social Science and Technology Award Catalog* published by the National Award Office.
   The above three categories of awards must be related to green mines construction to be accepted. We need to check the award management method for social science and technology awards, and the type of awards awarded must be consistent with the award management method.

Onsite evaluation or supporting materials:

The reward object must be in the unit; the superior administrative units, the same-level units, and subordinate units are not allowed.

68 R&D and technical renovation investment

Standard score: Six

Grading description: R&D and technological transformation investment shall not be less than 1.5% of the main business income in the previous year. Six points are rewarded for reaching 1.5%, five points for 1–1.5%, four points for 0.5–1%, and two points for those less than 0.5% and employees who invest in the technological innovation project

Assessment method: Search information

Basis or standard: Review subsidiary ledger or project ledger in financial statement

[Interpretation] This article requires that the investment in R&D and technical renovation shall not be less than 15% of the main business income.

1. R&D investment, technical renovation investment, and main business income
   R&D investment refers to the enterprise's expenses in the research and development of products, technologies, materials, processes, and standards.

   Technical transformation investment means that enterprises increase product output, improve product quality, increase product varieties, promote product upgrades, save energy, and reduce consumption and technological progress costs. The enterprises strengthen the comprehensive utilization of resources, the "three wastes" management, and labor safety. New technologies, new processes, new equipment, new materials are adopted to carry out technological renovation and renewal of existing facilities and process conditions and increase investment in technological measures.

   Main business income refers to the enterprise's business income from its industry's production and operation activities. The main business income of a mining enterprise refers to "product sales income." When an enterprise fills in the main business income, it generally fills in based on the cumulative number of the main business income indicators in the enterprise accounting "profit and loss statement" in the previous year. Only sufficient investment can ensure the gradual progress of R&D and technological transformation, the rapid transformation of scientific research results, and the acceleration of green mine construction.
2. Relevant regulations

   (1) The *Announcement of the State Administration of Taxation on Issues Concerning the Collection Scope of the Pre-tax Deduction of R&D Expenses* (State Administration of Taxation Announcement No. 40 of 2017) clarifies the scope of R&D investment expenses. *Notice of the Ministry of Finance and the State Administration of Taxation of the Ministry of Science and Technology on Improving the Policy for the Pre-tax Deduction of R&D Expenses* (Finance and tax [2015] No. 119) and *Notice of the Ministry of Finance, the State Administration of Taxation and the Ministry of Science and Technology on Increasing the Pre-tax Deduction Ratio of Research and Development Expenses* (Finance and tax [2018] No. 99) stipulate the applicable scope of additional deduction. At present, all enterprises can enjoy the preferential tax deduction policy of 75% before tax.

   (2) The relevant expenditures for enterprises' R&D activities shall comply with the *High-tech Fields Supported by the State* and *Guide to Priority Development of High-Tech Industrialization Key Fields* announced by the National Development and Reform Commission, the Ministry of Science and Technology, the Ministry of Industry and Information Technology, the Ministry of Commerce and the State Intellectual Property Office. Those who conform to the prescribed project's definition and scope can enjoy additional deduction of R&D expenses; those that do not confirm the

definition and scope will not be eligible for an additional deduction of R&D expenses.

(3) The main business income of a mining enterprise is the "product sales income." When an enterprise fills in the main business income, it is generally based on the cumulative number of the main business income indicators in the enterprise accounting "profit and loss statement" in the previous year.

3. R&D and technical renovation investment requirements

(1) Enterprises set up special R&D and technical renovation financial subjects Refer to Table 1 for the financial subjects of R&D and technical renovation.
(2) Special audits of scientific research or technological transformation shall be carried out.
(3) The investment in R&D and technological transformation shall not be less than 1.5% of the main business income in the previous year.
(4) The investment funds for mine R&D and technological transformation should be allocated in time according to the project plan to ensure the normal development of the project and promote the transformation of scientific research achievements and the construction of green mines

Onsite evaluation or supporting materials:

Financial statement.

69 High-tech enterprise qualification

Standard score: Three

Grading description: Three points are rewarded for obtaining the high-tech enterprise certification

Assessment method: Check the certificates

Basis or standard: Certifications of high-tech enterprises

[Interpretation] This article encourages enterprises to apply for high-tech technology enterprises.

High-tech enterprise refers to resident enterprises that have been registered in China (excluding Hong Kong, Macao, and Taiwan) for more than one year and that continues to carry out research, development, and transformation of technological achievements within the *High-tech Fields Supported by the State* to form the core independent intellectual property rights of the enterprise and carry out business activities on this basis.

In the *2018 Catalogue of High-tech Fields Supported by the State*, the green mine construction technologies have been listed in the category of " Chapter Seven Resources and Environment—Section 8 Resource Exploration, Efficient Mining and Comprehensive Utilization Technology—7. Green Mine Construction Technology".

Table 1 Reference for R&D and technical renovation financial subjects

Level1 subject	Level2 subject	Level3 subject	Level4 subject	Level5 subject	Level6 subject	Summarized content
R&D expenditures	R&D project	Expensed versus and capitalized expenditures	1. Own development 2. Cooperative development 3. Concentrated development	Personnel cost	Wage and salary	Wages, salaries, and overtime pay of the employees in the company involved in R&D activities
					Bonus subsidy	Allowances, subsidies, and bonuses for the employees of the enterprise participating in R&D activities, and year-end salary increases
					Others	Other related labor costs and other expenses of external R&D personnel
				Direct investment	Material fuel and power	Raw materials, semi-finished products, fuel and power costs (water, gas, and electricity) consumed by R&D activities

(continued)

Table 1 (continued)

Level1 subject	Level2 subject	Level3 subject	Level4 subject	Level5 subject	Level6 subject	Summarized content
					Test means purchase cost	Purchasing costs for molds, samples, prototypes, and general testing means that do not meet the fixed asset standards for intermediate tests and product trials
					Tooling and inspection fee	Process equipment development and manufacturing costs, equipment adjustment inspection fees, trial product inspection fees, etc
					Instrument and equipment maintenance fee	Simple maintenance fee for equipment used in R&D activities
					Equipment rental fee	Rental fees for instrument and equipment used in R&D activities
					Equipment purchase fee	Mainly only provide R&D equipment currently planned for use

(continued)

Table 1 (continued)

Level1 subject	Level2 subject	Level3 subject	Level4 subject	Level5 subject	Level6 subject	Summarized content
					Others	Other related expenses incurred in the implementation of research and housing rental fee for R&D activities
				Depreciation and long-term deferred	Instrument and equipment	Depreciation cost of instrument and equipment used for R&D activities
					Long-term deferred expense	Long-term deferred expenses incurred in the process of R&D equipment and instrument modification and repair
					Others	Depreciation expenses of buildings used in R&D projects, including long-term deferred expenses incurred in the process of reconstruction, modification, decoration, and repair of R&D facilities

(continued)

Table 1 (continued)

Level1 subject	Level2 subject	Level3 subject	Level4 subject	Level5 subject	Level6 subject	Summarized content
R&D expenditure	R&D project	Expense versus and capitalized expenditure	1. Own development 2. Cooperative development 3. Concentrated development	Design fee		Expenses incurred for the design, development, and manufacture of new products and new processes, design of procedures, technical specifications, and operational characteristics
				Equipment commissioning fee		Expenses in R&D activities in tooling preparation, including researching production machines, molds and tools, changing production and quality control procedures, or formulating new methods and standards
				Amortization of intangible assets	R&D software	Amortization of expenses incurred in R&D software purchased due to R&D activities

(continued)

Table 1 (continued)

Level1 subject	Level2 subject	Level3 subject	Level4 subject	Level5 subject	Level6 subject	Summarized content
					Patent right	Amortization of expenses incurred in the acquisition of patent rights due to R&D activities
					Generic invention(technology)	Amortization of expenses incurred for non-patent inventions (technology) purchased due to R&D activities
					Amortization of other intangible assets	Amortization of expenses incurred due to the purchase of licenses, proprietary technologies, design and calculation methods, and other special technologies required for R&D activities
				Other fees	Exploration field test fee	Onsite test fee for exploration and development technologies

(continued)

Table 1 (continued)

Level1 subject	Level2 subject	Level3 subject	Level4 subject	Level5 subject	Level6 subject	Summarized content
					R&D related expenses	Demonstration, appraisal, review, acceptance, evaluation of R&D achievements and application fees, registration fees, agency fees of intellectual property
					Library material and translation fee	Fee for technical book data and material translation
					Others	Meeting expenses, travel expenses, office expenses, foreign affairs expenses, R&D personnel training expenses, training expenses, expert consultation expenses, and high-tech R&D insurance expenses incurred for R&D activities
			4. Entrusting development			

The technologies included are "green mine design and construction technology, resource green mining technology, resource-efficient beneficiation technology, and mining area ecological and efficient restoration technology."

Onsite evaluation or supporting materials:

High-tech enterprise certification.

70 Intellectual property right
Standard score: Six

Grading description: Within three years, two points are rewarded for obtaining an invention patent, one point for publishing a core journal paper, and one point for a utility model or software copyright. All achievements should reflect the unit's name, and the total score should not exceed six points

Assessment method: Search information

Basis or standard: Original patent, monograph, or thesis

[Interpretation] This Article encourages enterprises to publish patents and papers.

1. Papers
   Foreign language journals, especially English journals, are divided into three categories: Science Citation Index Extended Edition (SCIE), Social Science Citation Index (SSCI), and Arts and Humanities Citation Index (AHCI). The three journal classifications overlap, in addition to engineering indexes (EI). The simplest way to determine which category the journal belongs to is to check the Web of Science website (http://mjl.clarivate.com/#opennewwindow) due to some disciplines' interdisciplinary characteristics.

   The level of Chinese journals is relatively simple, the basic standards of the liberal arts "Chinese Social Science Citation Index" (CSSCI) and the "Chinese Core Journals Overview"; the basic standard of science is the Chinese Science Citation Database (CSCD), now known as China Science and Technology Papers Statistics and Analysis System (CSTPCD).

   The core journal is the main journal of a certain discipline. It generally refers to professional journals with a large amount of professional information and high quality, representing the development level of professional disciplines and receiving readers' attention in the discipline.
2. Patents
   The patent refers to exclusive rights and benefits. The word "patent" comes from the Latin litterae patentes, which means open letters or public documents. It is a certificate used by medieval monarchs to issue certain privileges and later refers to a certificate of exclusive rights personally signed by the King of England.

   In modern times, a patent is generally a document issued by a government agency or a regional organization representing several countries based on an application. Such a document records the content of an invention-creation, and within a certain period, produces a legal state, that is, others can only exploit the invention-creation that has been patented with the authorization of the patentee. In China, patents are divided into three types: invention, utility, and appearance design.

Invention patents refer to a new technical proposal for products, methods, or improvements, emphasizing the technology.

The utility model refers to a new technical solution suitable for practical use proposed for the shape, structure, or combination of the product, emphasizing the patent's structure.

Appearance design refers to a new design that is aesthetically pleasing and suitable for industrial applications based on the shape, pattern, or combination of the product, as well as the combination of color, shape, and pattern.

3. Software copyrights
Computer software copyright refers to the various exclusive rights enjoyed by software developers or other rights holders following relevant copyright laws for software works.

Computer software copyright acquisition does not require individual confirmation, which is the referred principle of "automatic protection". After the software is registered, the software copyright owner shall enjoy the rights of publication, developer's identity, use, permission, and obtaining remuneration.

Onsite evaluation or supporting materials:

(1) Patent or computer software copyright certificate.
(2) Published papers.

71 Advanced technology and equipment

Standard score: 20

Grading description: Select mining and dressing techniques, technologies and equipment encouraged, supported and promoted by the state, and included in the "*Environmental Protection Technology Catalog Encouraged by the State*," "*Promotion Catalog of Advanced Applicable Technologies for Conservation and Comprehensive Utilization of Mineral Resources*," *National Advanced Pollution Prevention Demonstration Technology Directory*, "*Safe Production Advanced Applicable Technology, Process, Equipment and Material Promotion Catalog*," "*National Key Energy-saving Technology Promotion Catalog*," and "*Energy-saving Electromechanical Equipment (Product) Recommended Catalog*," with application certification provided. Ten points are rewarded for each technique, craft, or equipment, and the total score shall not exceed 20 points

Assessment method: Search information, and check the site

Basis or standard: Related industrial policy catalogs, design specifications, and related certification materials

[Interpretation] This article refers to the requirements for the processes, technologies, and equipment selected by mining enterprises.

The technologies and processes encouraged, supported, and promoted by the state refer to the technologies and processes in the catalog of advanced applicable technologies, processes, equipment, and materials issued by relevant state departments. The advanced applicable technology catalogs issued by relevant state departments include the *Catalogue of Advanced Applicable Technologies for Conservation and Comprehensive Utilization of Mineral Resources*, *Catalogue of Environmental Protection Technologies Encouraged by the State*, *Catalogue of National Advanced Pollution Prevention Demonstration Technologies*, *Promotion Catalog of Advanced Applicable Technologies, Process, Equipment and Material*, *Promotion Catalog of National Key Energy-saving Technology*, *Recommended Catalog of Energy-saving Electromechanical Equipment (Product)*, and *Industrial Structure Adjustment Guidance Catalog*.

1. Release units of advanced technology and equipment and construction significance.

   Table 2 shows the release units of advanced technology and equipment and construction significance.

2. Proof material of advanced applicable technology and equipment.

   (1) Declaration units in the catalog of advanced technology and equipment and relevant documents can be found in the catalog of relevant government websites.

   (2) The technology in the advanced technology and equipment catalog used, and certification provided by the technology application unit (the technology provider shall also provide relevant certification at the same time).

   (3) The parameter models listed in the advanced technology and equipment catalog are entirely consistent with those used by the enterprise, which can be verified from the scene.

Onsite evaluation or supporting materials:

Relevant supporting materials.

Table 2 Release units of advanced technology and equipment and construction significance

Content	Units	Significance
Catalog of Advanced Applicable Technologies for Conservation and Comprehensive Utilization of Mineral Resources	Ministry of natural resources	Promote advanced and applicable technologies for the conservation and comprehensive utilization of mineral resources
Catalog of Environmental Protection Technologies Encouraged by the State	Ministry of ecological environment	Accelerate the industrialization demonstration and promotion of energy-saving and emission reduction technologies, and guide environmental protection industries
Catalog of National Advanced Pollution Prevention Demonstration Technologies	Ministry of ecological environment	Speed up the demonstration, application, and promotion of advanced pollution prevention technologies for environmental protection
Promotion Catalog of Advanced Applicable Technologies, Process, Equipment and Material	Ministry of emergency management	Promote advanced and applicable technologies, processes, equipment, and materials for safe production
Promotion Catalog of National Key Energy-saving Technology	National development and reform commission	Popularize key energy-saving technologies, guide energy-consuming units to adopt advanced new energy-saving technologies, technologies, and equipment, and improve energy efficiency
Recommended Catalog of Energy-saving Electromechanical Equipment (Product)	Ministry of industry and information technology	Lead the transformation and upgrading of energy-saving electromechanical equipment (products)
Industrial Structure Adjustment Guidance Catalog	National development and reform commission	Guide the direction of industrial restructuring

2 Intelligent Mine

72 Construction program of intelligent mine

Standard score: Five

Grading description: Two points are rewarded for smart mine construction in the company's annual plan, and three points for implementation according to the plan

Assessment method: Search information, and check the site

Basis or standard: Annual Corporate Plan

[Interpretation] This article refers to the requirement for mining enterprises to formulate intelligent mining construction plans.

Mining enterprises shall have intelligent mine construction plans. It includes the construction, integration and construction plan, investment budget, equipment selection of the network platform, data storage, centralized control platform, automation subsystem, information management system, and other subsystems.

Onsite evaluation or supporting materials:

The plan for intelligent mine construction in the enterprise's annual plan.

73 Mine automation centralized control platform

Standard score: Ten

Grading description: Ten points are rewarded for meeting the requirement of building a centralized mine automation management and control platform, displaying the automatic control system, remote monitoring system, reserve management system, and monitoring systems

Assessment method: Check the site

Basis or standard: The construction plan of the mine automation centralized control system

[Interpretation] This article requires mining enterprises to build a digital centralized mining control platform.

The Mine automation centralized control system comprises integrated comprehensive monitoring, integrated communication, and integrated information management and control. Each subsystem's real-time data shall be in unified storage, intelligent management, real-time analysis, centralized monitoring, centralized release, unified management and control, and unified scheduling to realize the scientific prediction, planning, control, and decision-making command of mine development and operation. It can improve the mining enterprises' operational efficiency and safety production level, and the competitiveness of mining enterprises, and achieve sustainable development.

1. Integrated, comprehensive monitoring platform

    (1) Realize the remote monitoring and control functions of the main production systems, such as transportation, power supply, drainage, ventilation, and compressed air in the whole mine, to achieve unattended operation, reduce personnel and increase efficiency.

    (2) By integrating the whole mine's production system data, the intelligent linkage, intelligent alarm, and intelligent emergency plan management among related systems can be realized.

2. Integrated communication system platform
   Realize the integration of wired and wireless data transmission, dispatch communication, emergency broadcast, personnel positioning, industrial television, other systems of the whole mine, efficient integration of voice, video, personnel management, and emergency linkage to improve emergency dispatch and command.
3. Integrated information management and control
   (1) Realize the whole system's analysis application system, such as the early warning, detection, fault analysis, and operation report with the equipment as the object to reduce the incidence of failures and shorten the time of failure maintenance.
   (2) By integrating real-time information related to the production execution process, the production operation state can be reflected in time to realize the entire production's optimized management and the integrated management and control.
   (3) Take safety production management as the mainline, set the goal of enhancing the comprehensive application value of mine safety production information, cover the business of the main production functional departments of the mine, and provide information sharing and business collaboration for managers at all levels of the mine.

Onsite evaluation or supporting materials:

Digital mine planning scheme, centralized control software, and dispatching center.

74 Automation mine production system

Standard score: Ten

Grading description: ① Four points are awarded for establishing an unattended automation system for fixed facilities in central substations, water pump rooms, fan stations, air compressor rooms, and belt transportation lanes. ② Three points are awarded for establishing remote control systems for main equipment in the mining and production process. ③ Three points are awarded for establishing safety monitoring system platforms of waste rock yard, waste slag yard, slope construction, and working environment

Assessment method: Check the site

Basis or standard: Mine automation subsystem construction scheme

[Interpretation] This article requires the construction of a mine production automation system.

Build the mine production automation system to realize the centralized control and information linkage of production, monitoring, and control subsystems. By realizing the mechanization of equipment and automation of the system to reduce personnel replacement, increase work efficiency, and maintain mine production at the optimal level.

The basic elements of mine automation are the information and data acquisition system, high-speed and large-capacity two-way communication and information system network, computerized information management and automated mine planning, control and maintenance, and automatic and remote mechanical equipment connected with the information system network of the whole mine. The basic condition of mine automation is the control ability of real-time automation, changing and optimizing the internal factors continuously according to changes in the enterprise's internal and external conditions so that the output value reaches the highest level. The mine's comprehensive automation system realizes remote centralized monitoring and control of the whole mine's safe production and realizes the unattended underground work. The mine automation system mainly considers the following two factors.

(1) Establish an automated centralized management and control system for mines, focusing on the following automation systems: ① Unattended automation system for fixed facilities in a central substation, water pump room, fan station, air compressor room, belt conveyor lane, and other places. ② Safety monitoring system platforms for the waste rock yard, waste slag yard, and other storage yards and slope construction; online monitoring platforms for waste-gas and wastewater pollution control systems.
(2) Install a remote monitoring system to achieve centralized control and remote operation of remote equipment.

Onsite evaluation or supporting materials:

Planning scheme of mine production and automation system.

75 Remote video monitor and control system

Standard score: Ten

Grading description: Establish a complete remote video monitoring system. Remote video monitoring systems shall be installed in production sites such as mine working faces, key points (including power supply, drainage, ventilation, transportation, metering, and sales), and important safety sites (including tailings ponds and roadways). One point is awarded for each installation and real-time monitoring, with a total score of no more than ten points

Assessment method: Search information, and check the site

[Interpretation] This article requires installing video surveillance systems at key points in the mining area.

The remote monitoring system can monitor the equipment operation, manual operation, unsafe environment, key equipment, and personnel to ensure personnel safety and reduce safety accidents.

Critical workplaces mainly include production sites such as mine working face, key points in the process of power supply, drainage, ventilation, transportation, measurement, sales, and essential safety sites such as tailings ponds and roadways. Specifically, the main installation locations of remote video monitoring points should be cave mouths, car yards, unloading ports, winch houses, explosives depots, pump houses, substations, coal plants, tailings depots, dumps, weighing houses, and sewage treatment plants.

Onsite evaluation or supporting materials:

The main remote-video surveillance signals can be found in the dispatch center.

76 Resource reserve management system

Standard score: Five

Grading description: Five points are rewarded for the actual work of 3D reserves management

Assessment: Check the site

[Interpretation] This article requires enterprises to build 3D models of resource reserves and reserve management software.

The digital resource reserve model creates mine formation, ore model, structural model, or other models by establishing a geological database and triangulation modeling technology. According to the mining field's international block model concept, the grade model was established using geostatistical valuation methods. Establish a 3D spatial data visualization analysis model using the geological data obtained by various methods based on the 3D visualization platform, the rapid delineation technology, algorithm, and 3D geological body modeling method based on geological exploration data are studied.

The digital-resource economic model adopts the internationally recognized "block model + geostatistics" reserve calculation method, combining domestic reserve classification standards and traditional reserve calculation methods, as well as domestically recognized mineral economics principles and geo-economic indicators to develop highly integrated application software. This software conforms to the national reserve estimation standards and economic evaluation specifications. It integrates the original sample catalog, combination, probability distribution model, experimental variation function calculation, theoretical variation function fitting, the reliability cross-validation of the variation function model, block modeling, grade estimation, and reserve calculation. The rapid delineation and economic evaluation of the deposit's economic model can be realized with different technical and economic indicators. The main considerations are as follows.

(1) Establish a 3D model of resource reserves to realize the dynamic management of resource reserves.

(2) Establish a resource economic model to realize dynamic economic evaluation of resources.
(3) Construct the software management system of resources and reserves.

Onsite evaluation or supporting materials:

(1) There is a 3D reserve-management system.
(2) It can be operated on the dispatch center or other computers.

77 Intelligent working face or driverless mine-car system

Standard score: Five

Grading description: Five points are awarded for one of the following two items: ① Setting up the intelligent working face for normal production; ② Building a driverless mine car system

Assessment method: Search information, and check the site

Basis or standard: Intelligent working face or driverless mine car design scheme

[Interpretation] This article encourages coal mines to build the unmanned working surface.

1. Exploit the unmanned mining face in the coal mine
   Unmanned working-face mining in coal mines means that workers do not appear in the working face but operate and control electromechanical equipment outside the working face to complete various processes, such as coal breaking, coal loading, coal transportation, supporting and treating goaf in the working face. Unmanned face mining in the coal mine is an advanced and efficient stoping technology.

   The advantages of mining technology in the unmanned coal-mining face are as follows:

   (1) There is an important connection between the mining technology of the unmanned face in coal mines and the geological conditions. Based on the working face's complete system, wired or wireless methods are adopted to complete the information transmission and detection and control the key production equipment. The whole working process involves many specialties and technologies, such as autonomous positioning and navigation technology, coal and rock automatic identification technology, and automatic scraper conveyor moving technology. In the cooperation between sub-technologies, the mining technology automation of the unmanned working face in the coal mine has been realized, and the dynamic optimization of the entire operation process has been completed.

(2) High mining safety, high level of mining technology, and wide mining range of coal seam are the advantages of mining technology of the unmanned working face in coal mines. The mining technology of the unmanned working face in coal mine can complete the pre-release of gas in high-gas coal seams and control the working face's gas concentration within the safe range so that the safety of mining can be improved. Since the mining technology of the unmanned working face in coal mines has a wide range, the mining range includes thin coal seams, medium coal seams, thick coal seams, and extra-thick coal seams.

(3) The mining technology of the unmanned coal-mining face also provides an important basis for traditional coal seam mining technology and integrates advanced production technology. Through in-depth research on mining methods, mining equipment, and production organizations, it can use automatic control technology and new mining equipment to realize intelligent coal mining and effectively improve coal production efficiency.

2. Driverless mining trucks
   The driverless mining trucks adopt vehicle wire control technology, which can operate functions such as reversing, parking, automatic dumping, trajectory operation, and autonomous obstacle avoidance accurately and smoothly on the mine site. At the same time, with the continuous progress of digital intelligent mine construction, the driverless mining trucks can automatically load, transport, and unload without human operations following mine scheduling instructions by configuring environmental perception systems, behavior control and decision-making systems, positioning systems, and high-precision maps for mine vehicles. Finally, combined with the vehicle collaborative operation platform, the driverless vehicle fleet can operate cooperatively, with one vehicle perception, data sharing, and overall visibility, to achieve high efficiency and safety in mining operations.

Onsite evaluation or supporting materials:

Coal-mine unmanned working face or driverless mining trucks design scheme.

78 Mine environment online monitoring system

Standard score: Five

Grading description: Establish an online monitoring system for the mine environment with the functions of automatic grading alarm and grading notification for the pollutants (mine water, air pollutants, solid waste, and noise) regulated by the environmental protection administrative department following the law; five points are rewarded for meeting the requirements

Assessment method: Search information, and check the site

Basis or standard: The construction plan of environment online monitoring system in the mining area

[Interpretation] This article requires the construction of an online environmental monitoring system.

Environmental monitoring determines environmental pollution and environmental quality by monitoring and measuring indicators reflecting the environmental quality. The contents of environmental monitoring mainly include physical index monitoring, chemical index monitoring, and ecosystem monitoring.

(1) Water quality monitoring is divided into water environment quality monitoring and wastewater monitoring, and the former includes surface water and groundwater. The monitoring items include physical and chemical pollution indicators and relevant biological indicators, as well as hydrological parameters such as flow velocity and quantity of flow.
(2) Air detection is divided into air environmental quality monitoring and pollution source monitoring. Air monitoring often needs to determine the wind direction, wind speed, air temperature, air pressure, humidity, and other meteorological parameters.
(3) Soil monitoring: The key monitoring items are heavy metal elements, harmful nonmetallic elements, and residual organic pesticides affecting soil's ecological balance.
(4) Solid waste monitoring includes industrial wastes, wastes from health care institutions, agricultural wastes, radioactive solid wastes, and municipal solid wastes. The main monitoring items are the hazardous characteristics of solid waste and the characteristics of domestic waste, as well as the determination of the composition and content of toxic and hazardous substances and toxicology experiments.
(5) Biological monitoring and biological pollution monitoring: Biological monitoring uses biology to monitor environmental pollution. Biological pollution monitoring uses detection methods to monitor toxic and hazardous substances in organisms. The monitoring items are mainly heavy metal elements, harmful non-metal elements, pesticide residues, and other toxic compounds.
(6) Ecological monitoring: Observing and evaluating ecosystems' response to natural and human-made changes is a measurement of the temporal and spatial patterns of each ecosystem's structure and function, focusing on changes in biological communities and populations.
(7) Physical pollution monitoring refers to monitoring physical factors that cause environmental pollution, such as noise, vibration, electromagnetic radiation, and radioactivity.

Here is mainly for monitoring mine drainage, air pollutants, solid waste, noise, and other monitoring aspects.

Onsite evaluation or supporting materials:

The system experiences the functions of automatic grading alarm and classification notification.

Chapter 9
Enterprise Management and Enterprise Image

Enterprise management and enterprise culture reflect the enterprise's management level, which ensures the coordination of resource development and ecological protection from the system and the continuous improvement of green mine construction.

The enterprise management evaluation results should be reflected in the enterprises' management system's guarantee ability to ensure green mines construction. The evaluation process should consider the existence of systems and requirements and understand whether these systems and requirements are actually in place and are consistent with the continuous improvement principle of green mine construction.

1 Green-Mine Management System

The green mine management system formulates, implements, realizes, reviews, and maintains the organization, planning activities, responsibilities, practices, procedures, processes, and resources required by the green mine policy.

The green mine management system enables mining enterprises to achieve their green mine construction goals and continuously improves the mining environment, resources, comprehensive utilization, energy saving, emission reduction, intelligent mine construction, and technological innovation for a new level.

The green mine management system improves the relationship between mineral resources development and ecological protection, focusing on playing a leading role to enable all staff to participate in the green mine management, implement process control, and achieve continuous improvement.

79 Green mine construction plan and target
Standard score: Five.

J. Deng et al., *Interpretation of Green Mine Evaluation Index*,
https://doi.org/10.1007/978-981-16-5433-6_9

Grading description: Five points are awarded for green mine construction content, target, index, and corresponding measures in the annual enterprise plan.

Assessment method: Search information.

Basis or standard: Enterprise annual plan.

[Interpretation] This article requires the enterprise to plan the targets and indicators of green mines.

The organization shall establish and maintain green mine goals and indicators and form a file for each relevant function and level.

When an organization establishes and reviews green mine goals, it shall consider laws, regulations and other requirements, important green mine factors, alternative technical solutions, financial, operational and business requirements, and relevant parties' views.

The targets and indicators shall be consistent with the green mine policy and include a commitment to coordinate resource development and ecological protection. Simultaneously, the goal and index of green mines are to decompose and refine the green mine policy.

1. Targets, Indicators, and Measures of Green Mines

   Targets, indicators, and measures of green mines are the foundation of the whole green mine construction. Only by sorting out the targets, indicators, and measures can enterprises truly build green mines. Figure 1 shows the relationship in the green mine management system.

   First of all, the factors affecting green mine shall be determined. According to the factors affecting the construction of green mines, the targets and indicators of green mine construction are decomposed, and the green mine management plan is formulated according to the targets and indicators. The management plan can be purchasing a set of equipment, implementing an engineering project, or specific measures.

   Typical targets include soil and water conservation standards, mining greening standards, resource development and utilization standards, science and technology investment requirements, noise emission standards, dust emission standards, and clean and beautiful mining management.

   Common indicators include green coverage rate, disturbed land consolidation rate, forest and grass vegetation restoration rate, land reclamation rate, recovery rate, mineral processing recovery rate, comprehensive resource utilization rate, dilution rate, science and technology input, and dust concentration, noise.

   Green Mine Evaluation Index is an index in a broad sense and an experience requirement in a narrow sense.

   Table 1 shows the control objectives, control indicators, possible impacts, and green mine construction measures.
2. Green Mine Policy

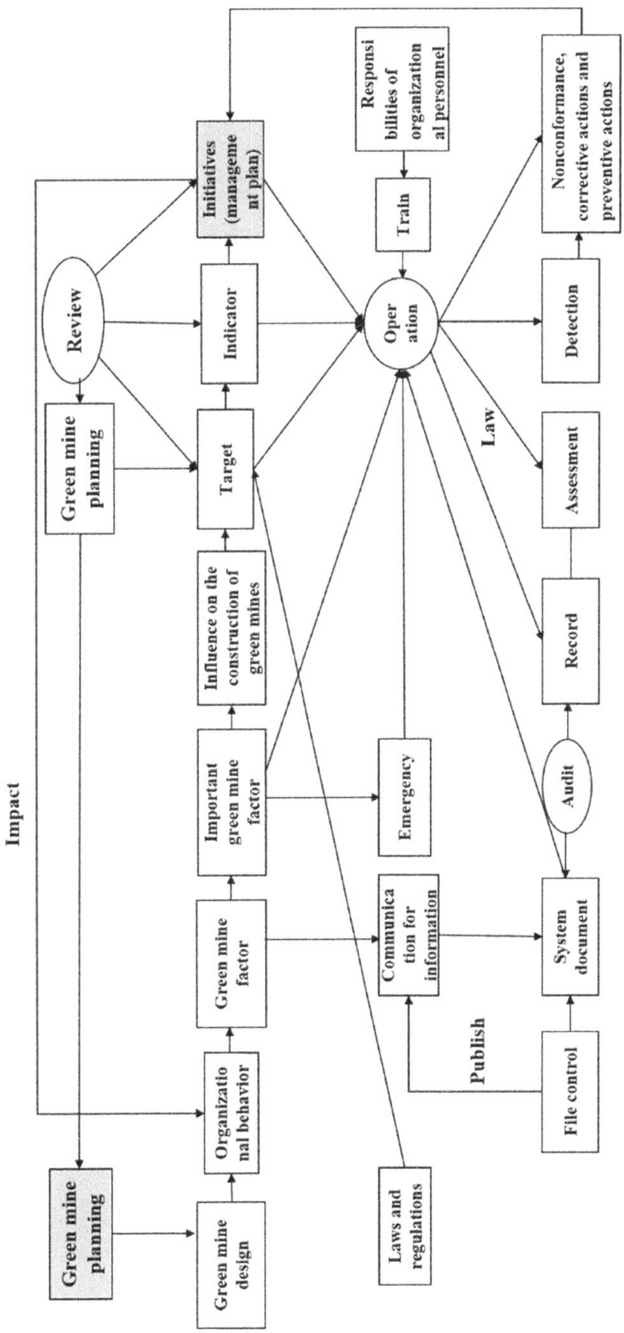

Fig. 1 Green-mine management system

Table 1 Green mine construction control targets, control indicators, possible impacts and measures

Serial number	Control targets	Control indicators	Possible impacts	Measures
1	No green area and no greenery	The green coverage rate of mining area reaches 100%	Without providing a comfortable environment for the staff	1. Establish a greening safeguard mechanism 2. Construct a greening system
2	Wastewater discharge compliance	*Integrated Wastewater Discharge Standard* *(GB 8978—1996)*	Pollute the surrounding environment	1. Construct the sewage treatment systems according to the types of pollutants 2. Recycle wastewater
3	Solid waste discharge compliance	*General Pollution Control Standards of Industrial Solid Waste Storage and Disposal Sites* *(GB 18,599—2001)*	Pollute the surrounding environment	1. Identify stacking sites according to pollutant discharge status 2. Construct pollutant stacking sites as required 3. Make corresponding environmental protection work for the pollutant stacking site 4. Manage the stacking site
4	Dust concentration at post compliance	*Occupational Exposure Limits for Harmful Factors in the Workplace* *(GBZ 2.1—2007)*	Pollute the surrounding environment and endanger people's health	1. Use rock drilling and dust collection integrated drilling rig for dust collection or wet rock drilling techniques for dust fall in rock drilling operations 2. Spray water for dust fall during the blasting operation 3.Set dust removal and dust collecting equipment at fixed dust-producing points, and keep enough negative pressure running synchronously to suppress and deal with dust generated in collection and processing

(continued)

Table 1 (continued)

Serial number	Control targets	Control indicators	Possible impacts	Measures
5	Noise at post compliance	The noise standard is 85 dB	Environmental pollution endangers people's health	1. Sound insulation building, sound insulation wall, sound insulation greenbelt 2. Carry out noise reduction treatment on high-noise equipment such as rock drilling, crushing, and air pressure in the mining area, equipped with noise reduction, shock absorption, and shock isolation measures
6	Noise emission at the factory boundary compliance	Sixty dB at day and 50 dB at night	Environmental pollution	1. Reasonable formulation and implementation in place according to the actual situation of the project 2. Establish and improve the environmental management responsibility system 3. Comply with the requirements of a civilized construction site 4. Implement pre-control measures 5. Implement safety/ environmental protection techniques disclosure for sub-projects
7	Domestic sewage discharge compliance	Domestic sewage discharge reaches the national secondary discharge standard	Environmental pollution	Set up a sedimentation tank and sewage treatment tank for discharge after treatment
8	Dust emission is unreasonable	Dust above 1.5 m cannot be detected visually during construction	Environmental pollution	Implement dust pollution prevention and control measures

(continued)

Table 1 (continued)

Serial number	Control targets	Control indicators	Possible impacts	Measures
9	Surface settlement control	No significant surface subsidence	Geologic, environmental damage	1. Adopt filling mining and water conservation mining 2. Reserve ore pillar or artificial replacement ore pillar
10	Working face specification	Step height, width, and slope angle meet the requirements	Cause geological disaster	1. Adopt millisecond blasting, pre-split blasting, and smooth blasting 2. Select mechanical equipment and the direction of working face advancement scientifically according to the occurrence of ore and geological engineering conditions 3. Strengthen the management of construction operations

The "green mine policy" refers to the organization's general green mine purpose and direction as officially released by its top management. Generally, the green mine policy is consistent with the organization's general policy and provides a framework for formulating green-mine building. It is precisely the commitment of the enterprise to society, responding to the responsibility of the enterprise.

The top management of mining enterprises shall determine the organization's green mine policy and ensure that it is within the green mine management system's scope. The green mine policy shall achieve the following points.

(1) Suitable for the nature and scale of organizational activities and their impact on resource development and ecological protection.
(2) Include a commitment to continuous improvement and maximizing the coordination between resource conservation and ecological protection.
(3) Include a commitment to applicable laws and regulations and other requirements related to green mines' construction.
(4) Provide a framework for establishing and evaluating green mine construction targets and indicators.
(5) Implement and maintain based on the formed documents.
(6) Convey to all those who work for or on behalf of the organization.
(7) Available to the public.

Proof of acceptance and scoring supplement:

It includes a table of targets, indicators, and measures for the main standard green-mine terms.

80 Green Mine Construction Organization and Responsibility

Standard score: Five.

Grading description: Five points are rewarded for having a clear organization and responsibility system for green mine construction.

Assessment method: Search information.

Basis or standard: Relevant documents on the establishment and responsibilities of green mine management institutions.

[Interpretation] This article requires enterprises to establish an organizational structure for green mine construction.

Roles, responsibilities, and powers shall be defined, documented, and communicated to facilitate green mines' effective management.

The manager shall provide necessary resources, including human resources and special skills, technical and financial resources, to implement and control the green mine management system.

The top management of the organization shall appoint a special manager representative (green mine commissioner). In addition to other responsibilities, they should also clarify their roles, responsibilities, and authorities in the following aspects:

(1) Ensure the establishment, implementation, and maintenance of the green mine management system requirements in accordance with the standards.
(2) Report the green-mine management system's performance to top management for review and provide evidence for improving the green mine management system.

Enterprises shall publish the organization structure and responsibilities of green mine construction with formal documents and responsibilities assigned to people.

Onsite evaluation or supporting materials:

For the organization structure, the responsibility documents of the enterprise's formal green mine construction should be sealed; it will not take effect without the seal.

81 Green Mine Assessment
Standard score: Five.

Grading description: Establish a green mine assessment mechanism and conduct internal assessments at least once a year against the green mine construction plan and targets. Five points are awarded for meeting the requirements.

Assessment method: Search information.

[Interpretation] This article refers to the requirement for assessing green mine targets and the completion of indicators.

The evaluation and assessment of the completion of the green mine construction plan or target determine the problems existing in the green mine construction process to achieve continuous improvement. In the process of green mine construction, green mine management documents shall be compiled so that that green mine construction can be standardized and orderly.

The green mine assessment here is an assessment of targets and indicators. Article 79 shows relevant targets and indicators.

1. Assessment Method

   (1) Self-assessment according to relevant requirements.
   Conduct self-assessment (self-inspection) at least once as required. Self-assessment for green mine declaration is not included.
   (2) Evaluate the performance of green mine construction according to the target and index form.
   The performance evaluation table should reflect the completion and improvement of each indicator with relevant instructions.

2. Basis Documents for Assessment

   (1) Green mine management system documents.

(2) Process data.

Relevant materials shall be retained in green-mine construction as an important basis for performance appraisal. These materials include notices of green mine management meetings, minutes of meetings on green mine management, purchasing records, engineering and other activities, and relevant archived materials.

3. Constitution of Green Mine Management System Documents

(1) *Green Mine Construction Manual*. It is a programmatic document and action code for green mines management in enterprises. It should clarify the green mine construction policy and describe the document of its green mine management system. It provides a systematic, specific, and programmatic description of the green mine management system.
(2) Program files. The supporting document of the *Green Mine Construction Manual* describes the implementation of the green mine management system. It provides the required methods for various functional departments' activities and links the Green Mine Construction Manual and the operation documents.
(3) Job files. The supporting documents of the program documents are the indicative documents given for specific job activities.
(4) Green mine construction program. Stipulate special green mine improvement measures, resources, and sequence of activities for a particular project or contract. For specific procurement, engineering, production, transportation, scientific research, and other projects, mining enterprises can separately prepare purchase, engineering, production, transportation, scientific research, and work out plans for green mines planning or green mine implementation rules.
(5) Green mine construction record. Documents are used to record the status of activities and the results achieved. In terms of green mine projects such as procurement, engineering, production, transportation, scientific research, and green mine construction, project records will form green mine construction documents.

Onsite evaluation or supporting materials:

(1) Green mine performance appraisal form or green mine performance evaluation content.
(2) Self-evaluation report (declaring self-evaluation report of the previous period).

82 Green Mine Construction Improving and Promoting

Standard score: Five.

Grading description: Five points are awarded for ensuring improvement content, measures, responsible person, completion time, and the effect achieved of green mine construction.

Assessment method: Search information.

[Interpretation] This article is a request for improvement.

Continuous improvement is the core process of strengthening the green mine management system of improving the overall performance of green mines according to the organization's green mine policy, which is also a plan based on a particular baseline. Table 2 shows the green mine improvement plan.

Onsite evaluation or supporting materials:

Improve the tables of measures.

83 Green Mine Construction Training

Standard score: Eight.

Grading description: ① One point is awarded for the green mine training system and plan. ② Three points are awarded for organizing managers and technicians to conduct green mine construction training (learning). ③ Four points are awarded for regularly organizing full-time green mine personnel to participate in green mine construction systematic training (learning) with training (learning) proof.

Assessment method: Search information, and perform a spot check on employees.

Basis or standard: Training system, training plan, training sign-in, video material, training notices, certificates, and photos.

[Interpretation] This article is a requirement for green mine training for relevant personnel of mining enterprises.

The training system for green mine employees means to standardize and tighten the training plans, requirements, implementation, and other aspects of enterprises in rules and regulations. The specific situation of each enterprise is different, and the training system is also different. Strengthen the training system for all employees in constructing green mines, establish a training account, and set up professional positions for green mines. Following the *Green Mine Construction Training Program*, train in a planned, post-differentiated, level-differentiated, and phase-differentiated manner.

1. Requirements for Training in the Management System

Table 2 Green mine improvement plan

First level indicators	Secondary indicators	Principal	Starting and ending time	Measure content	Bonus point	Cost/yuan
Mine environment	Mine appearance	Li	04/12/2020–05/14/2020	Improve the fixed plan, strengthen daily assessment, and the enterprise management department is responsible	2	10,000
Mine environment	Mine appearance	Zhang	03/26/2020–03/28/2020	There is an unauthorized built house in front of the old transportation office building without formalities, which must be demolished	3	100
Mine environment	Mine appearance	Qiao	04/12/2020–04/30/2020	Identify the location of missing signs and boundary signs, and ask relevant manufacturers for making them	1	5,000
Mine environment	Mine greening	Zhao	04/16/2020–04/30/2020	Construct automatic sprinkler irrigation for greening before office building	4	150,000
Mine environment	Mine greening	Li	04/13/2020–04/20/2020	The agency service team carries out the greening and planting work	4	500
Mine environment	Mine greening	Zhao	04/152020–05/15/2020	Install 100 m green isolation band from point A to point B	5	5,000
Resource development mode	Mine environment restoration and land reclamation	Mu	04/13/2020–06/26/2020	Conduct comprehensive treatment of slopes to eliminate safety hazards and reduce the difficulty of ecological restoration	10	50,000

(continued)

Table 2 (continued)

First level indicators	Secondary indicators	Principal	Starting and ending time	Measure content	Bonus point	Cost/yuan
Resource development mode	Environmental management and *monitoring*	Zhang	04/10/2020–04/30/2020	Apply for management system certification	5	500
Comprehensive utilization of resources	Solid waste resources utilization	Wu	04/06/2020–04/132020	Establish full tailings filling system to reduce ore dilution rate	24	450,000
Comprehensive utilization of resources	Domestic waste disposal	Li	04/102020–04/30/2020	Develop and implement the domestic waste classification management system	4	500
Energy-saving and emission reduction	Exhaust emission	Liu	03/19/2020–03/27/2020	Add the gas cap collection equipment to collect polluted gas, and treat it uniformly	8	500
Total					70	672,100

A procedure shall be established and maintained to make so that personnel at each relevant function and level are aware of:

(1) The importance of complying with green mine policies and procedures and complying with the green mine management system's requirements.
(2) The actual or potentially significant impacts of their work activities and the green mine benefits of individual job improvement.
(3) Their roles and responsibilities in implementing green mine policies and procedures and implementing the green mine management system requirements, including emergency preparedness and response requirements.
(4) Potential consequences of deviation from the prescribed operating procedures.

The staff who may significantly impact green mines shall have appropriate education, training, and work experience to be competent for their work.

2. Training Organization
The organization shall determine the training needs. All personnel whose work may have a significant impact on the green mine should be required to undergo appropriate training.

(1) Formulate training plans for green mine construction.
The training plan is a record arranged in a certain logical order. It is a pre-system setting of training content, training time, training location, trainer, training object, training methods, and training costs based on a comprehensive and objective analysis of training needs, starting from the organization's strategy.

Mining enterprises formulate training plans for green mine construction according to the requirements of personnel and actual conditions, and the relevant requirements and content of the "*Green Mine Construction Training Program*" to ensure that the staff in each position can meet the requirements of the green mine construction and complete the related tasks in terms of the green mine construction.

(2) Organize or participate in green mine construction training.
Organizational training means that the organization adopts various methods to organize personnel types for education and training investment activities to improve labor productivity and personal satisfaction with the occupation and serve its production and operation directly and effectively.

The main management personnel of mining enterprises and green mine construction specialists shall participate in the green mine training approved by the Ministry of Natural Resources and the local resources departments/bureaus, and understand the connotation of green mines, clear green mine construction industry standards, the meaning of each green mine evaluation indicators, and the target of green mine construction direction. At the same time, green mine specialists, or experts who participate in the compilation (or interpretation) of industry standards, or experts who prepare (or interpret) local standards, shall conduct internal training for employees.

The number of employee training each year, the location, number of people, person-number, and training contents shall be recorded clearly and effectively.

(3) Training and assessment.

It is recommended to obtain a relevant department's training certificate after participating in the assessment organized by the relevant authority to ensure training quality.

Onsite evaluation or supporting materials:

Training certificates, training plan, training notices, and training sign-in.

2 Enterprise Culture

The construction of enterprise culture is to condense people in thought, regulate people in behavior, and improve people in quality, thus creating a people-oriented management concept and development strategy for cultivating talents in the corporate management system to solve the problem of "people" in the development of the enterprise.

84 Survey on Employee Satisfaction

Standard score: Three.

Grading description: Carry out employee satisfaction questionnaire surveys regularly, set up questionnaire survey content reasonably, and be objective and fair. One point is awarded for organizing once a year, or the satisfaction level higher than 70%, or publishing timely.

Assessment method: Perform spot checks on employees.

Basis or standard: Original records of questionnaires.

[Interpretation] This article requires enterprises to carry out employee satisfaction surveys.

Standard requirements: Carry out employee satisfaction questionnaire surveys regularly and irregularly, and set up the questionnaire survey content reasonably to be objective and fair. The satisfaction degree of enterprise employee shall not be less than 70%.

The principle of employee satisfaction survey is an activity jointly carried out by the labor union (standard original text: the enterprise union organization shall be improved, and give full play to their role to enrich the material, sports, cultural life of the employees, and the employee satisfaction of the enterprise is not less than 70%) and the human resources department.

From enterprises, the employee satisfaction survey can understand the employees' specific problems in their daily work, the enterprise's work progress from another perspective, and the defects in the work that cannot be found at ordinary times. In the employee satisfaction questionnaire, some views and opinions about the enterprise that the employees do not usually express can also be found because the survey is generally anonymous. Therefore employees can feel free to fill in and express their reasonable suggestions.

From the employee's perspective, employees can see the enterprise's emphasis on employees' work from an employee satisfaction questionnaire, but they can also express their opinions about the enterprise, or some suggestions that they cannot express at ordinary times in the form. Of course, they can also express their demands for the company and make some suggestions regarding employee benefits and employee salaries when answering the questionnaire. Therefore, through employee satisfaction questionnaire surveys, employees can benefit and see the importance the enterprise attaches to their work.

Conducting employee satisfaction surveys is responsible for employees and a platform for enterprises to communicate with employees. Therefore, no matter in designing the questionnaire or answering the questionnaire, both parties' cooperation is required to achieve the ultimate purpose. The ultimate purpose of making an employee satisfaction questionnaire is also mutual benefit and win–win results.

1. Purpose of the Employee Satisfaction Survey

   (1) To improve employee satisfaction and loyalty, it is necessary to investigate employee satisfaction and improve the enterprise's management for sustainable development.
   (2) In order to grasp the mental dynamics and psychological needs of employees, targeted countermeasures should be taken to reflect the humanized care of the enterprise for employees and prevent the loss of talents.
   (3) Establish a formal communication mechanism for all employees to deal with the problems affecting employees' satisfaction timely and continuously enhance the centripetal force and cohesive force of employees to the enterprise.
   (4) Understand employees' opinions and views on the enterprise's management system and processes, working environment and atmosphere, work efficiency, communication and feedback, management leadership, work returns, human resources, logistics support, and other aspects of operation and management.

2. Specific Method of the Employee Satisfaction Survey

   (1) Establish an employee satisfaction survey institution (maybe temporary) to be responsible for the employee satisfaction survey.
   (2) Establish the *Employee Satisfaction Management System.*
   (3) The survey of employee satisfaction mainly includes the design and distribution of questionnaires, as well as the collection, analysis, and summary of questionnaires.
   (4) After the investigation, the *Employee Satisfaction Analysis Report* shall be compiled, and the compiler is responsible for classifying, counting, analyzing, judging, and discussing information in the employee satisfaction survey, to form an "*Employee Satisfaction Analysis Report*" with collective opinions.
   (5) Employee-satisfaction information-release methods include email, written form, bulletin board posting, and meeting.
   (6) The employee satisfaction survey is a general survey of all employees rather than a few employees' random check. The employee satisfaction survey result is an important indicator for measuring enterprise management work. Therefore, the questionnaire shall be carefully designed to play its due role.

Onsite evaluation or supporting materials:

(1) Check the labor union organization, *Worker Representative Conference System*, related records provided by mining enterprises, employee satisfaction survey forms, summary tables, survey reports and return ledger, and other supporting materials.
(2) Review the documents or compilation, consult and understand the employee satisfaction survey, and convene the employee representative meeting during the employee seminars and questionnaire surveys.

85 Recreational Activities of the Staff

Standard score: Four.

Grading description: ① Two points are awarded for staff leisure, entertainment, cultural, and sports facilities. ② Two points for facilities operating normally.

Assessment method: Search information, and check the site.

[Interpretation] This article requires enterprises to provide leisure, entertainment, cultural, and sports facilities for employees.

1. Functions of Recreational and Sports Activities of the Employees

   (1) As an important part of enterprise culture construction, employees' recreational and sports activities play a unique role in activating the working environment and improving employees' comprehensive quality. Due to

the different general employees' education levels, it is difficult for all employees to understand the true meaning of enterprise culture construction only through text introduction. Recreation and sports activities can convey the rigid, mechanical, and dogmatic parts of the enterprise culture through books, paintings, songs, dances to the employees in an entertaining way, making it easier for them to understand and accept.

(2) Carrying out employee cultural and sports activities can help relieve employees' fatigue after long-term tedious work and reduce employees' mental fatigue and behavioral inertia due to a single operation link. With the improvement of material and cultural living standards, employees are no longer satisfied with passive rest after work and expect to use collective entertainment to eliminate fatigue and ease work tension. Employees also hope to satisfy their interests through entertainment and release excessive energy and bad mood. Meanwhile, with the improvement of employees' cultural literacy, they appreciate literary and artistic works, organizing literary and activities, and studying calligraphy, painting, and photography as the pursuit and enjoyment of beauty. Colorful recreational and sports activities can create space for employees to display their talents and provide conditions for employees to cultivate their talents; they can meet their healthy and progressive spiritual needs, reflecting the "people-oriented" concept of enterprise culture.

(3) Carrying out employee cultural and sports activities can transform employees' consciousness behaviors into collective-consciousness behaviors. In daily work and behaviors, each person is a separate individual, and his thinking and way of thinking are disorderly and disorganized. The organization emphasizes more collectivist concepts and behaviors. If an organization lacks a collective concept, it will be in a loose and unstable state. Literary and artistic activities have the unity of purpose, the coordination of actions, overall cooperation, and information communication, which can drive the participants out of the scope of personal thinking and reach a collective consensus. Due to cultural and sports activities' collective nature, each employee should abandon narrow personal prejudices and establish collectivist ideas during participating in activities. Meanwhile, it revises employees' understanding and regulates their behaviors. The employees are encouraged to obey the collective goals, achieve harmony, and strengthen mutual cohesion. Activities such as basketball games, chorus, and tug-of-war competitions play an irreplaceable role in improving employees' awareness of collective honor and enhancing corporate cohesion.

(4) The development of cultural and sports activities for employees and the strengthening of corporate culture have complementary effects. Cultural and sports activities are an essential medium for promoting corporate culture and indirectly promote corporate culture. Simultaneously, the corporate culture broadens the space for cultural and sports activities, and its content and form show diversity and diversification. So the status

of cultural and sports activities in the hearts of most employees can be enhanced. In this way, while affecting employees' spirit and behavior, it provides great help for the construction of corporate culture and strengthens employees' cohesion and the competitiveness of the enterprise in the same industry.

2. Facilities for Staff Cultural and Sports Activities
   The leisure, entertainment, cultural, and sports facilities provided by mining enterprises for employees include massage chairs, saunas, fitness equipment, chess and cards, bathing, KTV halls and boxes, badminton halls, and basketball courts.

Onsite evaluation or supporting materials:

There are various leisure, entertainment, cultural and sports facilities for employees. The picture is only for reference.

Whether various leisure, entertainment, cultural and sports facilities for employees are in regular use.

86 Trade union organizations carry out activities

Standard score: Three

Grading description: Three points are rewarded for the trade union regularly carrying out activities to promote communication between employees and enterprises

Assessment method: Search information

[Interpretation] This article requires enterprise trade unions to carry out activities regularly.

1. Trade Union
   The trade union, the bridge and the link between the enterprise and employees, plays the following roles.
   (1) Safeguard the legitimate rights and interests of employees. For example, the general membership meeting or congress involves employees' vital interests. Labor remuneration, rest and vacation, labor safety and health, welfare, and other benefits are submitted to the conference for discussion to maintain the employee's rights for knowing, expression, and supervision.
   (2) Improve the quality of employees. Cooperate with the Ministry of Personnel and carry out skill training to promote the employees' career the production efficiency.

(3) Organize various technical competitions and labor competitions to promote the completion of production tasks and technological upgrading of enterprises.
(4) Organize employees to carry out cultural and sports activities, such as ball games, evening parties, and outreach training, to enrich their spare time and relax their minds and bodies.
(5) Assist employees who encounter difficulties, such as serious illnesses, natural disasters, and emergencies. The trade union will assist employees with some funds and call for other employees to help them overcome difficulties.

In short, if the enterprise union operates well, it will not cause trouble to the business. On the contrary, the enterprise's cohesion and employees' stability and happiness will be improved. It will create a win–win situation where the enterprise's efficiency and employees' income increase.

2. Trade Union Activities

(1) It should regularly or irregularly carry out employees' satisfaction questionnaire surveys and reasonably set up the questionnaire survey content. To be objective and fair, the enterprise's employee satisfaction should not be less than 70%.
(2) There are leisure, entertainment, cultural, and sports facilities for employees to ensure regular operation.
(3) The trade union regularly carries out activities to promote exchanges between employees and enterprises.

Onsite evaluation or supporting materials:

(1) Activity plan, activity record, and activity report.
(2) The notifications, videos, news, photos of ball games, evening parties, outreach training, and other activities every year.

87 Green-Mine Cultural Construction

Standard score: Three.

Grading description: There are propaganda films about green mine, based on the consideration of key content such as clarity, commentary, and duration. Three points are rewarded based on production effects as appropriate.

Assessment method: Watch promotional videos.

[Interpretation] This article requires enterprises to produce promotional videos to publicize the construction of green mines.

Compared with traditional publicity methods, the advantages of corporate videos are as follows: The promotional video contains a large amount of information, and the scene is authentic and credible. Video dissemination is impressive, which can be easily transmitted on the Internet. Thus, more and more industrial companies choose to customize promotional videos for companies. The main functions of the green mine propaganda film are as follows.

(1) The green mine promotional video allows target customers to more intuitively. The green mine construction policy, the green mine construction goal, the green mine construction practice, and the corporate spirit can be understood. The audio-visual combination method is more expressive and appealing than the traditional static picture. Customers can use a much easier way to understand the construction of the enterprise's green mines.
(2) The green mine construction promotional video can be easily transmitted through the network, enabling more employees to understand the enterprise's green-mine construction. It is also a form of "oversight through inspections by randomly selected inspectors of randomly selected entities and the public release of inspection results."

Onsite evaluation or supporting materials:

The enterprise provides promotional videos and contains relevant content on green mine construction.

3 Enterprise Management

88 The Linkage Mechanism Between Employee Income and Corporate Performance
Standard score: Two.

Scoring description: A mechanism to increase employees' income in line with the enterprise's performance should be established. Two points are awarded if the enterprise's total income is related to the enterprise's economic efficiency growth.

Evaluation method: Search information, check employees at random.

Basis or standard: Assessment system.

[Interpretation] This article requires that the employee's total income be linked to the enterprise's economic efficiency growth.

The *Opinions of the State Council on Reforming the Wage Determination Mechanism of State-owned Enterprises* (No.16 Document of State Council. [2018]) proposes to reform the method for determining total wages. Following the national macro policy requirements on wage income distribution, enterprise development strategy and salary strategy, annual production and operation goals, and economic benefits, enterprises should comprehensively consider labor productivity increase, labor cost input and output rate, and market benchmarking of employee wage levels. Combined with the salary guidelines issued by functional government departments, the total annual salary is reasonably determined.

Enterprises should establish a good performance salary distribution plan to reflect the mechanism of enterprise employees" income growth in pace with enterprise performance and realize the linkage between wages and benefits. If the economic benefits increase, the increase in the total wages of the year can be determined within the range that does not exceed the increase in economic benefits. Among them, if labor productivity does not increase in the current year, the labor cost input–output rate of the previous year is lower than the industry average, or the average wage of employees in the previous year is significantly higher than the average wage of employees in urban units across the country, the increase in total wages for the year should be lower than that in economic benefits over the same period. For enterprises whose main business is not in fully competitive industries and fields, the average wage of employees in the previous year has reached the level of regulation and above specified by the functional government departments. The increase in total wages for the year should be lower than that in economic benefits over the same period, and the increase in average wages of employees should not exceed the wage growth targets set by the functional government departments.

1. Establish an "income-Increase System for Enterprises and Employees"

   Enterprises should establish a "system of increasing income between enterprises and employees" or hold annual meetings of collective contracts on salary and forming minutes.
2. Establish a Salary Distribution Plan

   Enterprises should establish a good performance salary distribution plan and reflect how enterprise employees' income grows in pace with enterprise performance.
3. Synchronous Growth Mechanism of Performance and Revenue

   (1) The socialist labor distribution system is reflected—distribution according to work and making a concerted effort to create good results. The total income of employees is related to the enterprise's economic benefit growth.

   (2) Reasonably set enterprise assessment targets and indicators and make complete performance rewards and promotion mechanism.

Adopting proof materials and scoring supplement:

(1) Documents reflecting the relationship between employees' total income and the growth of the enterprise's economic benefits, such as financial statements and tax payment certificates.

(2) Implementation status: relevant implementation documents (such as assessment and bonus distribution) and other supporting materials.

89 Functional Area Management System

Standard score: Two.

Grading description: Two points are rewarded for having the management system of production and living in line with the enterprise's actual situation, and clarifying the responsible units or departments.

Assessment method: Search information.

Basis or standard: Check relevant management documents of mines.

[Interpretation] This article refers to management requirements for each functional area.

Each functional area has a corresponding management system, and the responsibility is assigned to the relevant unit or department.

(1) The management focus of each functional area should include: Item placement, hygiene, greening, vehicle parking, garbage sorting, waste stacking, dormitory, canteen, bathhouse hygiene, work dress, and management of related disciplines. It includes fixed management and visual management.

(2) There is no graffiti on the wall. There is no uncivilized content or content inconsistent with the party's guiding spirit and the State.

The functional area management system should combine fixed management and visual management.

Adopting proof materials and scoring supplement:

The management system is sound and in line with the actual situation of the mine.

90 Mining and dressing equipment management

Standard points: 20

Grading instructions: (a) There is a list of core equipment, including equipment name, model, main parameters, energy consumption, purchase time, and maintenance. (b) The onsite inspection equipment matches the list and can be used, and there is no outdated production process equipment explicitly eliminated by the country. Five points are rewarded for each item

Assessment method: Search information

Basis or standard: Check relevant management documents of mines

[Interpretation] This article is a requirement for the management of core equipment for mining and dressing.

The core equipment in this article refers to the high energy consumption in the main mining and selection process. It produces dust and noise has an impact on the environment. The main large-scale equipment of the mine refers to excavators, loaders, grab loaders, mucking machines, loaders, patio (shaft) drilling rigs, heading machines, coal mining machines, coal cutters, coal plough, comprehensive excavators, scraper conveyor, hydraulic supports, reversed loader, conveyor, drilling machinery, oil extraction machinery, workover machinery, crushing machinery, grinding machinery, screening machinery, sorting machinery, dewatering machinery, and sand-making machinery.

1. Main Mining Equipment in an Open-Pit Mine
   (1) Piercing equipment: impact drills, down-the-hole drills, and roller drills.
   (2) Mining and loading equipment: excavators (multi-bucket and single-bucket), bucket shovel, and front-end loaders. Single-bucket excavators are widely used.
   (3) Transportation equipment: automobiles and conveyors.
   (4) Dumping equipment: bulldozing plough, bulldozer, front-loading machine, drag scraper, and dragline.
2. Main Mining Equipment in Underground Mines
   (1) Metal mines: Scrapers, rock drilling rigs, crushers, and drilling rigs.
   (2) Coal mine equipment: Comprehensive coal-mining machine, comprehensive excavators, drill jambo, scraper conveyor, hydraulic supports, and belt conveyor.
3. Mineral Processing Equipment
   Crushers, soil-removal sieve, belt conveyor, shapers, and feeders.
4. Reference table of core equipment management of mining and dressing.
   See Table 3 for core equipment management of mining and dressing.
5. Outdated Production Process Equipment Explicitly Eliminated by the State
   The outdated production technology and equipment explicitly eliminated by the State are mainly *the Industrial Structure Adjustment Guidance Catalog* issued in 2005, 2011, and 2013. *The Industrial Structure Adjustment Guidance Catalogue (2019 Edition)* released in 2019 is the latest catalog. A total of 48 industries are involved, with 1,477 entries. Among them, 821 are encouraged, 215 are restricted, and 441 are eliminated.

Onsite evaluation or supporting materials:

Equipment List.

91 Occupational health management system

Standard score: Three

Grading description: Three points are rewarded for possessing occupational health management systems

Assessment method: Search information

Basis or standard: Check relevant management documents of mines

Table 3 Reference table for management of selected core equipment

Number	Equipment Name	Model	Parameter	Purchase time	Power	Maintenance	Current location
1	#1Jaw Crusher	C6X125	Maximum feed lumpiness 900 mm $B =$ 175 mm, Q = 540t/h	September 6th, 2018	320	Daily maintenance	Rough workshop
2	#1 draught fan	4–72-No.5A(4)	$Q =$ 11054m^3/h, $P =$ 2962 Pa $n =$ 2900 rpm, $N = 15$ kW	February 6th, 2019	30	Daily maintenance	Rough workshop
3	Drifting jumbo	DD311	Total motor power 70 kW, Drilling depth 4.2 m	April 5th, 2019	70	Daily maintenance	Working face
4	Scraper	ACY-2	Bucket volume 2m^3	November 2nd, 2018	50	Daily maintenance	Working face

[Interpretation] This article is about the management of occupational health.

The occupational health management system includes occupational health management objectives, occupational health management institutions, occupational health management facilities, and occupational health protection measures.

Onsite evaluation or supporting materials:

Occupational health management system and system documents should reflect the document version number and control status.

92 Environmental protection management system

Standard score: Three

Grading description: Three points are rewarded for having an environmental protection management system (including sewage and wastewater discharge; solid-waste classification, stacking, and control; noise control; dust control)

Assessment method: Search information

Basis or standard: Check relevant management documents of mines

[Interpretation] This article is a requirement for mining enterprises to establish environmental protection systems.

Environmental management mainly includes sewage and wastewater discharge management, waste management, noise control, dust control, and light pollution control.

Onsite evaluation or supporting materials:

Environmental protection management system and system documents should reflect the document version number and control status.

93 Personnel visual management

Standard score: Four

Grading instructions: a) Internal employees enter the production site, uniformly wear labor protection clothing, which shall meet safety requirements. b) External personnel for visiting or inspection, learners, and contractor employees enter the production site and dress in accordance with the safety of the production site

One point is deducted for one person who fails to meet the requirements

Assessment method: Check the site

Basis or standard: Personnel visualized management system

[Interpretation] This article is a requirement for the visual management of personnel in mining enterprises.

Visual management organizes onsite production activities using visual perception information with intuitive images and suitable colors. It is a management method to improve labor productivity, and a scientific method of management using vision.

Visual management characteristics are as follows: Visual signal display is taken as the basic means to ensure it is visible to everyone. With the basic principles of openness and transparency, managers' requirements and intentions should be visible to everyone as much as possible, thus promoting autonomous management, or called autonomous control. Onsite operators can visually display their suggestions, results, and orders, and communicate with leaders, colleagues, and workers.

(1) Personnel visual management: The internal employees enter the production site, uniformly wearing labor insurance clothing. Outsiders (visitors, inspectors, learners, contractors, and employees) enter the production site. Their dress codes should meet the production site's safety requirements and shall be different from the internal staff.
(2) Visualized management of tools: Tools such as compressed gas cylinders, scaffolding, hand-held power tools, and electrician tools shall be checked regularly to confirm that they are intact, with qualified labels pasted in the obvious positions.

(3) Visual management of equipment and facilities: The apparent parts of equipment and facilities are marked with names and numbers, and pipelines and valves are colored uniformly and standardized. The medium names and flows are marked on the pipelines, and the control buttons, switches, and display names are marked on the instrument control panel and indicating devices. Dangerous equipment and facilities have warning information, with safety operation notice signs set. The utensils containing hazardous chemicals should be placed in categories, with signs set. The signs include necessary information such as chemical names, main hazards, and safety precautions. The signs and labels should be kept tidy, clear, and complete, without omissions, discoloration, fading, shedding, or incompleteness.
(4) Visual management of production operation areas: Warning signs of Red, Yellow and Green, No Entry without Permission and safety warning signs of Turn off Cell Phones, No Fireworks, and Poisonous Danger should be suspended near the entrance in inflammable and explosive places such as oil and gas gathering (field) stations, oil and gas storage areas, oil and gas chemical production areas, natural gas purification plants as well as toxic and harmful production areas. Safety evacuation diagram, occupational hazards warning signs (type, physical and chemical properties, concentration or intensity, hazard safety, and protective measures), the risk source diagram, and safety board for admission should be set up near the main entrance. All safety warning signs and occupational hazard warning signs shall be suspended in production areas or dangerous parts. The signs and labels shall be kept neat, clear, and complete without omissions, discoloration, fading, shedding, or incompleteness.

The meaning of the visual management implementation is to create an efficient and refreshing workplace and eliminate the onsite "three harms" (occupational injury, defects, and breakdowns) caused by the "three noes" (no harm to yourself, no harm to others, and no harm from others). Therefore, it is necessary to set a series of projects for onsite improvement activities.

Onsite evaluation or supporting materials:

(1) Visual management system.
(2) System implementation records, such as punishment records, photos, and videos.

94 Green mine publicity activities

Standard score: Six

Grading instructions: Publish positive reports in the media, and carry out publicity activities related to the construction of green mines, reports, and competitions in publicity weeks. Two points are awarded for each category, with a maximum of six points

Evaluation method: Search information and check the site

[Interpretation] This article is a requirement for green mine publicity activities.

Green mine publicity activities include publishing positive reports in the media and carrying out publicity reports, competitions, and publicity week activities.

1. Positive publicity
   Positive publicity is an effective way of uniting people and national strength and creating a positive atmosphere to stimulate vitality. The promotion of spiritual civilization, good ideological and moral cultivation, the cultural and ethical quality of the whole nation, competitiveness, and social morality can create an excellent social environment for the next generation's healthy growth.
2. Preach
   Preach refers to propaganda activities related to green mine construction organized by mining enterprise leaders who will invite industry experts, typical green mining entrepreneurs.
3. Competition
   The competition includes various forms of competition activities such as green mine network knowledge contest, answering questions onsite, 'and performance contest, enhancing the employees' understanding and awareness of green mines.
4. Publicity week
   Publicity week refers to carrying out green mine publicity activities from various angles, in an all-around way during a fixed period.

Onsite evaluation or supporting materials:

Proof of activities, newspapers, etc.

95 Employees' physical examination

Standard score: Four

Scoring description: Two points are rewarded when enterprises organize all employees to regularly take an annual physical examination. Two points are rewarded when the enterprises formulate physical examination plans and items in categories and establish occupational health surveillance

Assessment method: Search information

Basis or standard: Physical examination files

[Interpretation] This article is a requirement for employees' physical examination.

The employees' physical examinations are divided into entry physical examination and health checkup. The health examinations shall be based on the employees'' actual situation to formulate a medical plan, the checkup items, and checkup frequency, for example, traditional health examination (general personnel, traditional items), professional function tests (such as vision, hearing, heart, lung, endocrine, and other professional function checkups), special medical problem detection (such as hypertension, chronic diseases, and occupational diseases).

1. Entry Medical Examinations
   Ensure that the recruited employees' physical conditions are suitable for professional work through physical examinations. They will not cause the epidemic of infectious diseases in collective life and will not affect others due to their physical reasons. The entry medical examination has relatively fixed medical examination items and medical examination standards, and a professional medical examination center can guarantee the quality of the medical examination. The physical entry examination is responsible for enterprises and employees.
2. Health Checkup
   A health checkup is one of the most direct ways to make employees feel solicitude. Meanwhile, it makes them feel at home in the unit.

   (1) Increasing employees' work efficiency: The enterprise should reduce work incoordination caused by sickness and absenteeism, which affects the work progress. It should care about employees' health, and improve employees' sense of belonging, work enthusiasm, and work efficiency.
   (2) Reducing human resources loss: Enterprise should reduce diseases, sick leave, health leave, and corresponding man-hours to ensure regular operation.
   (3) Reflecting the enterprise's human care for employees: The enterprise should provide medical and healthcare consultations and services for employees to enjoy professional and personalized health guidance. The incidence of employees' morbidity and sub-health status can be reduced through health intervention measures, showing that the enterprise cares about employees. The enterprise should provide special health management plans for senior managers to protect the core resources. Regular physical examinations can be used to encourage employees to implement the enterprise's care for employees.
   (4) Effective control and prevention of occupational diseases: Through health examinations, occupational diseases of employees and employees of special types of work can be detected in time, so that effective control and prevention can be performed.
   (5) Reducing enterprise healthcare-related expenditures: This service has made a systematic and comprehensive consideration of enterprise's healthcare-related expenditures, making overall arrangements, and reducing ' 'total medical expenditures.
   (6) Employee training and promotion plan should be considered based on employees' physical health.

3. Establishing a Medical Examination System for Occupational Diseases in Mines
   Mining enterprises shall establish a *Mine Occupational Disease Physical Examination System* and standardize the management of occupational health.
4. Establishing Occupational Health Surveillance Files
   Occupational health surveillance files refer to the systematic inspection and analysis of the health conditions of persons exposed to occupational hazards for

prevention to find necessary measures for early health damage. The occupational health surveillance file contains the following contents.

(1) Basic personal information.
(2) Test results of occupational hazards in the workplace.
(3) The results of previous occupational health inspections and handling information.
(4) Occupational health physical examination report, occupational disease diagnosis and treatment, and other health materials.
(5) Other relevant materials that need to be stored in occupational health surveillance files.

Onsite evaluation or supporting materials:

(1) The certification materials for the employees' entry physical examination.
(2) The certification materials for the employees' health checkup.
(3) Formulate physical examination plans by classification (including physical examination items and the frequency of physical examinations).
(4) Occupational health surveillance files.

4 Community Harmony

96 Mine harmony

Standard score: Five

Grading instructions: Mining enterprises shall establish a good relationship with the township (street) and village (community), and settle disputes and contradictions on time

Evaluation method: Check employees at random or visit the community

[Interpretation] This article is a requirement for the occurrence of no conflicts.

With economic development and the continuous improvement of people's awareness of autonomy, practical contradictions and interest disputes exist objectively. For the same incident, people have different opinions. Meanwhile, they oppose the other side's views, thus forming conflicts. When the conflict escalates to petitions, demonstrations, and parades, it is called a conflict incident.

Mining enterprises shall establish consultation and negotiation mechanisms with the towns (sub-districts) and villages (communities) where the mines are located and handle various interest disputes in time. Consultation and negotiation mechanisms should be established, and green mines should be integrated into social security to resolve interest disputes in time and ensure green construction. The stakeholder participation mechanism is a guarantee for obtaining feedback to handle conflicts.

Mining enterprises should actively resolve conflicts and prevent conflicts from escalating:

(1) The principle of prevention first and prevention before it happens.
It is necessary to establish a sound social benefit coordination mechanism and an early warning mechanism for social stability. Follow early detection, early reporting, early control, and early resolution, and timely eliminate factors inducing mass incidents.
(2) The principle of handling affairs according to law and policies. It is necessary to strictly implement national policies to avoid mass incidents caused by ineffective and insufficient implementation.
(3) Principle of educating and preventing intensification.

It is necessary to target different groups and guide the masses to express their interest requirements rationally and legally. The enterprise should resolve interest contradictions and prevent conflicts from intensifying and expanding.

Onsite evaluation or supporting materials:

(1) The mining enterprise shall issue a letter of undertaking not to petition, demonstrate, or march within nearly one year. The assessors visit the masses on site.
(2) The assessors do not find petitions, demonstrations, marches, and other events from the Internet and other media.

97 Poverty alleviation or charity fundraising activities

Standard score: Five

Grading description: The enterprise regularly or irregularly carries out poverty alleviation or charity fundraising activities. Five points are rewarded for those who have carried out poverty alleviation or charity fundraising activities in the past two years

Evaluation method: Search information, and check employees at random

Basis or standard: Poverty alleviation contract or donation contract or relevant bills

[Interpretation] This article is a regulation for mining enterprises to carry out poverty alleviation or charity fundraising activities.

1. Donation Based on Charity
The *Law of the People's Republic of China on Public Welfare Donations* stipulates that only public welfare social organizations and non-profit institutions can receive donations in accordance with the law. The money and materials raised in public welfare fundraising activities organized by enterprises should be received by public welfare social organizations and non-profit institutions.

2. Poverty Alleviation to Help Poor Households Obtain Legitimate Rights and Interests and Remove the Burden of Poverty
    (1) Poverty alleviation has short-term and long-term plans and clear goals. Moreover, specific plans, steps, and measures are formulated for the planning requirements. The symptoms and the root causes should be combined organically, focusing on the root cause.
    (2) Poverty alleviation helps poor households solve their living difficulties through developing production, and poor areas develop their economy. Follow the path of getting rich through hard work to get rid of poverty.

Onsite evaluation or supporting materials:

(1) Donation needs an invoice issued by the receiving unit.
(2) Poverty alleviation requires a poverty alleviation plan and related expenditure invoices.

5 Enterprise Integrity

98 The situation of enterprises paying taxes according to law
Standard score: Four
Grading instructions: Enterprises actively pay taxes in accordance with the law and good faith. In tax evasion, two points are deducted every time it is found until four points are deducted
Evaluation method: Survey and interview, and search information
Basis or Standard: Certificate of the tax department

[Interpretation] This article is a requirement for enterprises to pay taxes according to law.

Enterprises should insist on paying taxes in full and register and go through tax payment procedures in accordance with the tax law. Enterprises can use their tax credit rating to prove their tax status according to law.

The tax credit rating refers to the tax authority's assessment of the taxpayer's tax credit in a certain period based on the taxpayer's performance of tax obligations.

The *Administrative Measures on Tax Credit (Trial)* shall come into force as of October 1st, 2014, and the *Trial Measures on the Administration of Tax Credit Ratings* (No. 92 Document of State Tax [2003]) promulgated by the State Administration of Taxation on July 17th, 2003 shall be repealed simultaneously.

In 2018, Announcement No. 8 (2018) of the State Administration of Taxation of *Announcement on Matters Concerning Tax Credit Evaluation* proposed to add a level-M tax credit rating. In November 2019, the State Administration of Taxation issued *the Announcement of the State Administration of Taxation on Matters Concerning Tax Credit Restoration* (State Administration of Taxation Announcement No. 37 of 2019). From January 1st, 2020, tax credit restoration shall be carried out by making credit commitments and correcting dishonest acts for corporate taxpayers included in tax credit management. Taxpayers are encouraged and guided to increase their awareness of taxation in accordance with the law and integrity, with a credit-based new tax supervision mechanism established. The credit ratings at all levels are as follows.

1. Level-A Credit
   Those with an annual evaluation index score of 90 points or more are considered as level A. However, one of the following conditions shall not be recognized:
   (1) The actual production and operation period is less than three years.
   (2) The tax credit evaluation result of the previous evaluation year is level D.
   (3) Irregular reasons for zero declaration or negative declaration of value-added tax or business tax within an evaluation year for consecutive three or six months in total.
   (4) The enterprise cannot set up the account book in accordance with the national unified accounting system and cannot provide accurate tax information to tax authorities based on legal and valid vouchers.
2. Level-B Credit
   If the annual evaluation index score is more than 70 points, but less than 90 points, it is level B.
3. Level-M Credit
   The following enterprises that have not committed any dishonest acts listed in Article 20 of the Trial *Implementation of Tax Credit Management Measures* are eligible for Level-M tax credit:
   (1) Newly-established enterprises.
   (2) Enterprises with no income from production and operation during the evaluation year, and annual evaluation index score of 70 points or more.
4. Level-C Credit
   If the annual evaluation index score is more than 40 points, but less than 70 points, it is level C.
5. Level-D Credit
   If the annual evaluation index score is less than 40 points or directly judged to determine the level, it is D-level. If any of the following situations occurs, the score will not be evaluated, and the level will be D:
   (1) Evading tax payment and the recovery of overdue tax, defrauding of an export tax refund, and falsely making out special invoices for value-added tax, which constitute tax-related crimes after judgment.
   (2) There is an act listed in the preceding paragraph that does not constitute a crime, but the amount of tax evasion (evasion of tax payment) is

more than 100,000 yuan and accounts for more than 10% of the total tax payable of each tax. Alternatively, tax violations include evading payment of overdue tax, fraudulently obtaining export tax rebates, falsely issuing special value-added tax invoices, etc., and having paid taxes, overdue fines, and penalties.

(3) Failing to pay taxes, overdue fines, and penalties in accordance with the conclusions of the tax authorities within the prescribed time limit;
(4) Refusing to pay taxes or refusing or obstructing the tax authorities from carrying out tax inspection and law enforcement according to law using violence or threat;
(5) Violation of provisions on the administration of value-added tax invoices or other provisions on the administration of invoices, which results in other units or individuals failing to pay, underpaying, or defrauding tax.
(6) Providing false declaration materials to enjoy preferential tax policies.
(7) Fraudulently obtaining tax refund for export from the State, and the export tax rebate (exemption) qualification has not expired;
(8) Having records of abnormal households or the direct liability-holders from the abnormal households register or operate;
(9) Registered or operated by the directly responsible person of the D-level taxpayer;
(10) Having other severe untrustworthy circumstances recognized by the tax authority according to law.

Admissible, evidentiary materials and scoring supplement:

(1) The enterprise should pay taxes according to law. One can query on "Inquiry of Enterprise Tax Credit Rating" in each province or the "Tax Information Evaluation Results Query" system. The tax credit-ratings refer to Levels A, B, and M.
(2) The enterprise shall sign labor contracts with all employees and pay their social insurance according to law. Insured units can inquire from human resources and social security network and print out their insurance payment certificates.

Note: The enterprise cannot inquire from the "National Enterprise Credit Information Disclosure System." It cannot issue a certificate from natural resources or other related entities; the monthly tax payment certificate is insufficient to prove tax payment according to law.

99 Enterprise's performance of relevant obligations

Standard score: Four

Grading instructions: ① The enterprise submits geological data as required. ② The enterprise submits the basic table of mineral resources statistics on time. Two points are deducted for each non-conformance

Assessment method: Search information

[Interpretation] This article is a provision for enterprises to perform social responsibilities.

The relevant obligations fulfilled by enterprises include social responsibility, submitting relevant information as required by the administrative department, and the publicity of exploration and mining information by mineral rights holders.

1. Social Responsibility
   The mining enterprise should fulfill its social responsibilities and insist on honesty and trustworthiness, which shall be considered mainly from the following aspects.
   (1) Mining enterprises shall seriously perform the contract system in business dealings.
   (2) In terms of financial management, the mining enterprise should strictly implement the national financial management system and perform financial accounting in accordance with the national unified accounting system. It should achieve true and complete accounting information without off-book accounts, with a normal settlement with the bank and no lousy record.
   (3) The mining enterprise should implement the *Safety Production Law of the People's Republic of China* to realize safe production without accidents.
   (4) The mining enterprise should strictly abide by labor security laws and regulations and establish a sound labor contract system.
   (5) The enterprise's social responsibility and image should be enhanced, equally important as the brand. The enterprise shall insist on integrity and be law-abiding and ensure that the products are genuine at a fair price.
   (6) The enterprise should persist in resource-conservation and sustainable development and actively assume responsibility to protect employees' health and benefits.
2. Submit Relevant Information as Required by the Administrative Department
   (1) Geological data collection system.
   The competent department of Land and Resources under the State Council and the competent departments of Land and Resources under the people's governments of provinces, autonomous regions and municipalities directly under the Central Government shall, following the law, exercise the administrative functions entrusted by the State for the unified management of geological data. The summarization of geological data is performed in accordance with the *Regulations on Geological Data Management*.
   (2) Basic table of mineral resources statistics.
   (1) Extracts of relevant provisions of laws and regulations.
   Article 21 of the *Administrative Measures for the Registration and Statistics of Mineral Resources* (Order No. 23 of the Ministry of

Land and Resources) implemented in 2004 stipulates: Under the provisions of Article 21, mining right holders, who fail to fill in the basic mineral resources statistical table, make false reports, conceal, refuse, and make late reports on mineral resource statistics, or refuse to accept the inspection and spot-checking or practice fraud, should be punished in accordance with the measures for Article 18 of *the Administration of the Registration of Mineral Resources Exploitation* and Article 16 of *Provisions on the Administration of Collection of Compensation Fees for Mineral Resources* as well as *the Statistics Law of the People's Republic of China* and its detailed rules for the implementation of the relevant provisions.

Article 41 of *the Statistical Law of the People's Republic of China* stipulates: Enterprises or institutions or other organizations in one of the following circumstances may be fined 50,000 yuan and be punished. If the circumstances are serious, they shall be fined more than 50,000 yuan and less than 200,000 yuan. Individual businesses may concurrently impose a fine less than 10,000 yuan: (1) Refusing to provide statistical data or failing to provide statistical data on time after urging; (2) Providing untrue or incomplete statistical data.

Article 6 of *the Provisions on the Punishment of Statistics Violations and Disciplinary Acts*: …Where the departments concerned and their staff members commit any of the following acts in the course of carrying out a statistical investigation, the persons responsible shall be given a warning, a demerit record, or a severe demerit record; if the circumstances are relatively serious, a demotion shall be imposed: (1) Deliberately delaying or refusing to report statistical data. (2) Failing to carry out investigation and verification while knowing that the statistical data are untrue, thus causing adverse consequences.

(2) Basic requirements for filling in a form.

The basic statistical table of mineral resources shall take the mining area delimited by the mining license as the basic statistical unit and shall be filled in by the mining owner. Regardless of its production scale, it must be reported separately; where a mining right owner (mining enterprise) operates more than one mine (where one person has more than one mining right), he must fill in one (right) certificate (mining license) form separately. If one mining right holder (mining enterprise) opens more than one mine (one person with multiple rights), one (right) certificate (mining license) and one table must be filled in separately.

Mine submission time: Before the end of January each year.

Admissible, evidentiary materials and scoring supplement:

Geological-data collection and the filling materials of mineral-resource statistical basic forms.

100 Information publicity

Standard score: Two

Scoring description: The enterprise will obtain two points for publicizing mining information for mining rights holders in accordance with regulations

Evaluation method: Check the information disclosure system for exploration and exploitation by mining owners

[Interpretation] This article requires enterprises to fulfill the obligation of public disclosure of exploration and mining information for mining rights holders.

1. Related Regulations
   Please refer to the Measures for the Disclosure of Prospecting and Mining Information by Mining Right Holders (Trial) for specific requirements.
2. Check the Website
   The query website is the Ministry of Natural Resources National Mining Rights Holders' Exploration and Mining Information Publicity System (http://kyqgs.mnr.gov.cn/). The mining right holder shall fill in the information on the provincial department's portal website in charge of natural resources where the exploration item or the mine is located.

Adopting proof materials and scoring supplement:

By inputting the mining owner's name or license number in the "Mining-Owner information Disclosure System," the mining owner's information disclosure situation can be inquired, and the content of the disclosure can be checked to see whether it conforms to the relevant requirements of the *Measures for Mining Owner Information Disclosure (Trial)*.

Chapter 10
Related Knowledge

1 Reference Documents

Implementation Opinions on Accelerating the Construction of Green Mines

Ministry of Land and Resources, Ministry of Finance, Ministry of Environmental Protection, General Administration of Quality Supervision, Inspection and Quarantine

China Banking Regulatory Commission, China Securities Regulatory Commission

Implementation opinions on accelerating the construction of green mines

Regulations of Land and Resources (2017) No. 4

This Notice shall apply to the administrative departments of land and resources of all provinces, autonomous regions, and municipalities, the competent departments of finance and environmental protection, Bureau of Quality and Technical Supervision (department of market supervision and management), all banking regulatory bureaus, securities regulatory bureaus and industry associations, China Geological Survey and other units directly under the Ministry of Land and Resources, as well as all departments and bureaus of the Ministry of Land and Resources.

The implementation opinions are formulated to comprehensively implement the decisions and arrangements of the *Opinions of the Central Committee of the Communist Party of China and the State Council on Accelerating the Construction of Ecological Civilization* (issued by the Central Committee of the Communist Party of China (2015) No. 12) and the *Outline of the 13th Five-Year Plan for National Economic and Social Development of the People's Republic of China*, to promote the implementation of the national mineral resources planning and the ecological civilization construction in the mining field, and accelerate the transformation of the mining industry and green development.

J. Deng et al., *Interpretation of Green Mine Evaluation Index*,
https://doi.org/10.1007/978-981-16-5433-6_10

1. General requirements

(1) Guiding ideology. We must comprehensively implement the spirits of the 18th National Congress of the Communist Party of China (CPC) and the 3rd, 4th, 5th and 6th Plenary Sessions of the 18th Central Committee of the CPC, and thoroughly implement the spirit of the General Secretary Xi Jinping's series of important speeches. Great efforts should be made to promote the "Five-in-One" master plan and coordinately promote the "Four-Prolonged Comprehensive Strategy". We need to firmly establish and thoroughly implement the concept of innovative, coordinated, green, open, and shared development. It is essential to adapt to the new normal of leading economic development, as well as a conscientious implementation of the decisions and deployments of the CPC Central Committee and the State Council on the construction of ecological civilization. Equally important is the need to adhere to the work orientation of "due diligence in protecting land resources, saving and utilizing land resources intensively and doing our utmost to safeguard the masses' rights and interests" and focus closely on the overall requirements of ecological civilization construction. Through government guidance, enterprise entities, leading standards, policy support, innovating mechanisms, strengthening supervision, fulfilling responsibilities, and stimulating vitality, the concept of green development is integrated into the entire process of mineral resources planning, exploration, development, utilization, and protection, to improve the transformation and upgrading of traditional mining, and the quality and efficiency of mining development.

(2) Overall goals

Construct a working mechanism of departmental cooperation and four-level joint creation, increase policy support, accelerate the construction of green mines, and form a new mining development model that meets ecological civilization construction requirements by 2020.

A new pattern of green mine construction should be formed. All newly-built mines should meet the requirements of green mine construction, and the transformation and upgrading of producing mines should be accelerated to meet the requirements gradually. We should establish a green mine model that is led by technologies and driven by innovation, implement hundreds of demonstrations of green exploration projects, build more than 50 green mining development demonstration areas, and form several new models, mechanisms, and systems that can be replicated and popularized.

Construct a new way to transform the development model of the mining industry. Persist in the coordination of mode transformation and growth stabilization, and innovate the new industrial development mode of resources intensive utilization and recycling and the new way of mining economic growth. Accelerate the upgrading of green environmental protection technologies and equipment and strengthen the mine ecological environment's comprehensive management. The intensive utilization of lands and conservation of cultivated lands in mining areas should be promoted to guide effective mining investment. The endogenous power for mining enterprises' green development shall be stimulated for the sustainable and healthy development of China's mining industry.

Establish a new working mechanism for the development of the green mining industry. Adhere to the mutual promotion of green transformation and management reform, and establish a green mine construction working system featuring joint innovation at the national, provincial, city, and county levels, enterprise-based construction, third-party evaluation, and social supervision. Besides, relevant departments shall improve the standard system of green exploration and green mine construction and the supporting incentive policy system to construct a long-term green mining development mechanism.

2. Establishing leading standards and building green mines

(3) Adjusting measures to local conditions and improving standards. Firstly, all localities should combine actual conditions and follow the green mine construction (See the Appendix) to refine and form local standards for green mines in line with the regions' actual situations. Secondly, all localities should clarify the assessment indicator requirements of green mine construction, including mine environmental features, development and utilization modes, intensive utilization of resources, modern construction of mines, harmony between mines and lands, and corporate cultures and images. A green mine standard system with national, industrial, local, and group standards should be established, covering all major industries with characteristics.

(4) Classifying guidance and gradually reaching the standards. In transferring new mining rights, it is necessary to clarify several relevant requirements and liabilities for breach of contract in the transfer contract following the green mine construction requirements and related standards. The requirements and liabilities include development methods, resource utilization, mine geological, environmental protection, governance and restoration, land reclamation, etc. The

planning, design, construction, and operational management of new mines should be promoted in accordance with the requirements of green mine standards. Regarding producing mines, all localities should make comprehensive deployments and requirements based on actual conditions, differentiate situations, actively promote the upgrading and transformation of mines, and gradually meet the green mine construction requirements.

(5) Setting examples for demonstration and advancing the whole. Select the cities or counties with remarkable green mine construction progress to build some green mining development demonstration areas. Promote the innovation of technological systems, standard systems, industrial modes, management methods, and policy mechanisms to explore and solve the critical problems, such as layout optimization, structural adjustment, protection, conservation and comprehensive utilization of resources, and overall planning of ground and underground. Improve the system of mineral resources planning, exploration, development, utilization and protection, and the incentive policy system for green mining development to create a good investment and development environment. Promote the construction of green mines across all regions to form the green mining development model areas. The model areas shall have reasonable layouts, intensive and high efficiency, suitable environments, harmony between mines and lands, and the regional economy's benign developments.

(6) Ecological priority and green exploration. Under the premise of adhering to ecological protection first and respecting the masses' will, optimize the layout of strategic actions for mine exploration breakthrough. Establish the concept of green environmental protective exploration, and implement the ecological environment protection measures for exploration construction to carry out the legal and green exploration earnestly. Develop and popularize new technologies and methods, such as airborne geophysical exploration and remote sensing. Accelerate the revision of technical standards and specifications for geological exploration and improve the standard green exploration techniques system. Trenching and other exploration methods significantly affecting the surface environment shall be moderately adjusted or replaced to reduce geological exploration's impact on the ecological environment.

3. Increasing policy support and speeding up the construction process

(7) Implementing mineral resources support policies. Regarding the mining indexes and mining rights of mineral species under the total amount control, if green mining enterprises conform to the

national industrial policy, priority should be given to green mines and green mining development demonstration areas.

The mining rights, meeting agreement transfer conditions, shall be granted priority to be transferred to green mining enterprises in the form of agreement.

(8) Ensuring the lands for green mine construction. In the adjustment and improvement of the overall land-use planning, all localities should include the project lands required for the construction of green mines into the overall planning arrangement, and give priority in the annual land-use plan to ensure the reasonable new construction land demands for new construction, renovation, and expansion of green mines.

For the mining lands, after completing the legal formalities for construction land, the lands can be transferred, leased, or leased first and transferred afterward. If the land is provided through transfer, under the premise of not exceeding the legal maximum transfer term, the land user can flexibly choose the term of land-use right transfer and carry out the transfer based on factors such as mine production cycle and exploitation life. The land user may also agree to pay the land transfer price in installments in the land transfer contract.

The green mining enterprises should be supported to reclaim and revitalize the existing industrial and mining land and link it with the new construction land. The construction of green mining development demonstration areas should be promoted in an integrated manner with the reclamation and utilization of industrial and mining wastelands, the rehabilitation of mine geological environment, the treatment of soil pollution in mining areas, and the land remediation. Related pilot and supporting policies are applicable. Under the premise of meeting the requirements of planning and ecology, it is allowed to use the cultivated lands increased from the reclamation of historical industrial and mining wastelands for the balance of cultivated land occupation and compensation.

For the agricultural land or other land damaged and unrecoverable caused by legal mine exploitation, the field investigation shall be carried out according to land change investigation requirements and procedures. It shall be included in the annual change survey after being reviewed and approved by the special report. If the cultivated land is involved, the amount of cultivated land shall be reduced according to the facts. However, the upper limit of the number of local controls must not be exceeded, and the basic farmland must be supplemented.

(9) Increasing fiscal and tax policy support. When arranging the funds for geological and mineral investigation and evaluation, based on

perfecting current fund management measures, the Ministry of Finance and the Ministry of Land and Resources shall appropriately support the eligible projects of green mining development demonstration areas.

While using the central funds, local governments can make overall arrangements for the funds of geology and mineral resources, mine ecological environment treatment, heavy metal pollution prevention, and land reclamation, give priority to supporting eligible projects in the green mining development demonstration areas, and gather funds to promote the transformation of mining-development mode and the mining environment. Efforts should be made to actively promote the sustainable economic and social development of mining areas and actively coordinate local financial funds. An incentive system should be established to reward outstanding green mining enterprises.

Within the scope of *High Technology Field Supported by the State*, enterprises developing green mining construction technologies and transformation of achievements and are qualified as high-tech enterprises, and the enterprise income taxes can be legally levied at a reduced rate of 15%.

(10) Innovating green financial support policies. Under the premise of strengthening the environmental, health, safety, and social risk assessments and management of investment projects in the mining field, the banking financial institutions are encouraged to develop the green mine characteristic credit products in line with the actual regional conditions. Under the principles of controllable risks and sustainable business, green mining enterprises' financial support to restore the environment, prevent heavy metal pollution, and recycle resources should be increased.

Financial institutions are encouraged to provide financial services and support for green mining enterprises that have complete environmental, health, safety, and social risk management systems, timely information disclosure, good interaction with stakeholders, purchase of environmental pollution liability insurance, and competitive, marketable and profitable products.

Encourage provincial governments to establish green mine project libraries and strengthen support for green credit. Incorporate green mine information into the enterprise credit system as an essential reference for banks to handle credit business and other financial services.

Support governmental guarantee agencies to explore the establishment of structured green mining guarantee funds to provide credit enhancement services for green mining enterprises and projects. Encourage social capital to set up various green mining industry funds to provide support for green mining projects.

Promote qualified green mining enterprises to list on the Small & Medium Enterprise (SME) board, Growth Enterprises Market (GEM) board, and mainboard in China, and list on the "New Three Board" and regional equity market for financing.

4. Innovating evaluation mechanism and strengthening supervision and management.

(11) Construction of enterprise and standard warehousing. Mining enterprises that have completed the green mine construction tasks or met the requirements and relevant green mine construction standards shall conduct self-assessments and submit assessment reports to the municipal and county-level land and resources departments. Relevant departments of land resources and environmental protection of cities and counties should entrust a third party to carry out onsite verification in the form of government purchase services. Those that meet the requirements of green mine construction shall be reported to appropriate provincial authorities level by level and included in the national green mine list. They are all open to the public and subject to supervision through the green mining development service platform. Green mining enterprises included in the list automatically can enjoy the relevant preferential policies.

(12) Social supervision and punishments for dishonesty. Green mining enterprises should take the initiative to accept social supervision, establish a complaint-response mechanism for major environmental, health, safety, and social risk events, and promptly accept and respond to the demands of local people, social groups, and other stakeholders. Following the requirements of "double random and one open," provincial departments of land and resources, finance, environmental protection, and other relevant departments shall conduct random inspections on the mines included in the list of green mines from time to time. The relevant municipal and county-level departments shall do an excellent job in daily supervision and management. The Ministry of Land and Resources, together with relevant departments such as finance and environmental protection, should regularly evaluate the construction of green mines in all provinces (autonomous regions and municipalities). The mines that do not meet the requirements of green mine construction and relevant standards shall be removed from the list and exposed publicly and shall not enjoy various supporting policies such as mineral resources, land, and finance. If the green mine construction task in the mining right transfer contract is not fulfilled, the relevant mining

right approval departments shall promptly investigate the relevant liabilities for breach of contract in accordance with the regulations.

5. Implementing the division of responsibilities and promoting the overall planning coordinately.

(13) Division of labor and cooperation for joint promotion. The Ministry of Land and Resources, the Ministry of Finance, the Ministry of Environmental Protection, and the General Administration of Quality Supervision, Inspection and Quarantine, with other relevant departments, shall be responsible for the development's overall arrangements of green mining. The development direction, policy guidance, targets, and construction requirements shall be defined, and the work guidance, organization, coordination, supervision, and inspection of all provinces (autonomous regions and municipalities) shall be strengthened. Departments and institutions of land and resources, finance, environmental protection, quality supervision, banking supervision, and securities supervision at all levels shall follow the unified leadership of the people's government at the corresponding level, work together to form a joint force, and accelerate the construction of green mines according to the division of responsibilities.

With relevant finance departments, environmental protection, and quality supervision, the provincial departments of land and resources shall be responsible for the organization and promotion of the green mining development in their respective provinces (autonomous regions and municipalities). Efforts should be made to formulate work plans especially, identify green exploration and demonstration projects, formulate local standards for green mining construction, improve the system of technical standards for green mines in significant industries, define supporting policies and measures, and organize cities and counties to accelerate green mining exploration and construction. According to the work layout requirements of the Ministry of Land and Resources and other departments, relevant departments shall optimize the green mining development demonstration areas, guide the corresponding municipal and county establishment of the construction work plan, and do an excellent job in the organization, promotion, supervision, and management. Before the end of December each year, related progress and effectiveness and supervision and inspection should be reported to the Ministry of Land and Resources and other departments.

Under the leadership of the people's governments at the corresponding levels, the relevant departments of land and resources, finance, and environmental protection at the municipal and county levels shall be responsible for the specific implementation. Specific work measures shall be proposed in strict accordance with work plans. Relevant departments should urge the mining enterprises to carry out green exploration, build green mines, and do an excellent job in daily supervision and administration.

Besides strengthening the Technical Committee for Standardization's guidance, China Mining Association and other industry associations and enterprises should be encouraged to participate in the research and formulation of green mine standards; thus, the national and industrial standards for green mines can be summarized and formed gradually.

(14) Award and compensation incentive, demonstration, and guidance. The competent departments of land and resources and finance should establish the incentive systems to reward the green mines that have achieved remarkable results. The Ministry of Land and Resources and the Ministry of Finance will work with relevant departments to select a certain number of excellent green mines from the national list of green mines every year to give praise and reward and play a demonstrative and leading role.

(15) Building a platform for publicity and promotion. Build a green mining development service platform on the portal website of the Ministry of Land and Resources to publish the green mining policy information, national green mines list, green mines and catalog and standards of green exploration technologies and equipment, and to publicize the green mining progress and typical experience in various regions. Give full play to the bridging role of China Mining Association and other industry associations, and strengthen industrial self-discipline. Scientific research institutes and consulting institutions should be encouraged to participate in the construction of green mines jointly. The information sharing, publicity, and promotion should be strengthened.

The present implementation opinions shall come into force from the issuance date and be valid for five years.

Ministry of Land and Resources, Ministry of Finance, Ministry of Environmental Protection, General Administration of Quality Supervision, Inspection and Quarantine

China Banking Regulatory Commission, China Securities Regulatory Commission

March 22, 2017

2. Notice on the Selection of Green Mines in 2020

Notice of the General Office of the Ministry of Natural Resources on the Selection of Green Mines in 2020

To the natural resources competent departments of all provinces, autonomous regions, and municipalities,

The relevant matters concerning the green mine selection in 2020 are hereby notified as follows to promote green mines' construction continuously and orderly.

1. Selection principles

    (1) Selection basis. There are nine industry standards, including the *Implementation Opinions on Accelerating the Construction of Green Mines* (Regulations of Land and Resources (2017) No. 4) and the *Construction Specifications of Green Mine in the Non-metallic Mine Industry*.

    (2) Selection scope. Independent mines (including oil and gas) with valid mining licenses and regular operations have not been subject to administrative penalties from the departments of natural resources and ecological environment in the past three years or received administrative penalties but have been rectified. It is not included in the list of abnormalities in the mining rights holders' exploration and mining information publicity system during selection. The mining area does not involve any nature reserves, and the mining period for the remaining reserves shall not be less than three years.

    (3) Quantity of selection. The typicality and representativeness of mining enterprises should be highlighted, and the number of selections from each province (autonomous region, municipality) shall not exceed 10 in principle.

2. Selection Procedures

    (1) Online application. Starting from June 1, 2020, mining enterprises shall be organized to log into the National Green Mines List Management Information System (hereinafter referred to as the List System) and fill in the relevant application information.

    (2) Self-assessments of mines. Organize the mining enterprises that have applied online to conduct self-assessments according to the requirements of green mine construction and industry standards, form self-assessment reports, and fill them in the List System.

    (3) Third-party assessments. In the form of government purchase services, third-party assessment agencies should be entrusted to

carry out onsite assessments of mines, and the third-party assessment reports should be formed according to the requirements of unified assessment indicators.

(4) Review of materials. Review the self-assessment reports, third-party assessment reports, and other materials.

(5) Onsite verification. Onsite verification shall be carried out on all mines that have passed the third-party assessments using open and secret inspections and data reviews.

(6) Publicity. Through the Internet, newspapers, and other channels, the selection and recommendation results shall be publicized within the province (autonomous region, municipality).

3. Work requirements

(1) Elaborate organization. All localities should strengthen the organizational leadership and financial support, and coordinate the selection and recommendation work during COVID-19 prevention and control. Being realistic, open, fair, and just, local relevant departments shall implement the work procedures strictly and unify the assessment standards. The self-assessment reports and third-party assessments shall be all filled in the List System without submitting the paper version to reduce the burden of enterprises and material submission.

(2) Strengthening the responsibility mechanism of third-party assessment agencies. All localities should strengthen the supervision and promote the standardization of assessment. The assessment agencies should be the enterprises, public institutions, social organizations, or recognized professional assessment agencies with independent legal entity qualifications. The assessment agencies and experts shall remain independent from mining enterprises and shall not prepare the mine self-assessment reports or have any business dealings with mining enterprises within one year before and after the assessments. It is strictly forbidden to collect fees from mining enterprises in any form and use assessments for illegitimate interests. If there are serious violations such as practicing fraud, playing favoritism, and committing irregularities, notices of criticism should be circulated, and the assessment reports should not be accepted.

(3) Strict spot checks. All localities should do an excellent job in daily supervision and carry out spot checks on the mines that have been included in the national green mine list according to the requirements of "double random and one open." The mines that do not meet the standards should be reported in time and removed from the List System. Relevant work can be carried out simultaneously with the onsite verification of the mining rights holders' exploration

and mining information publicity, play the role of social supervision, and gradually build a green mines credit system.

All provinces (autonomous regions and municipalities) are requested to complete the selection before September 30, 2020, and inform the Department of Mineral Resources Protection and Supervision of the Ministry about the selection and recommendation list and onsite spot checks. According to epidemic prevention and control, the Ministry will carry out the reviews through spot, online, video, and photo checks. Mines that meet the relevant standards will be included in the national green mines list after being announced publicly.

Contact person and telephone number: Yongchao Sun 010—66558137
Fax and E-mail: 010—66558277
ycsun@mail.mnr.gov.cn

General Office of the Ministry of Natural Resources
May 14th, 2020

3. Requirements for the Third-Party Assessment of Green Mine Selection

Requirements for the Third-party Assessment of Green Mine Selection

The third-party assessment of green mine selection refers to government administration departments' behavior in government purchase services to entrust relevant assessment agencies to assess the mining enterprises' compliance status of green mine construction and form the third-party assessment reports. To ensure the selection quality of green mines and standardize the third-party assessment, the requirements are formulated according to the *Implementation opinions on accelerating the construction of green mines* (regulations of Land and Resources (2017) No. 4) and the *Notice on the selection of green mines in 2020* (Official Letter of General Office of the Ministry of Natural Resources (2020) No. 839).

1. Assessment agency

    (1) The third-party assessment agency shall be an enterprise, public institution, social organization, or recognized professional assessment agency registered in the territory of the People's Republic of China with independent legal entity qualification. The ability to assess green mine construction is essential.
    (2) The competent department of natural resources shall disclose the content and procedure requirements of the third-party assessments to the public, select the third-party assessment agencies based on the actual work of local government purchase services, and define

the rights and obligations of the third-party assessment agencies through contracts, agreements, and other forms.

(3) The third-party assessment agencies and experts must remain independent of mining enterprises. They must not be associated with mining enterprises, participate in preparing the mine self-assessment reports, or have any business dealings with mining enterprises one year before and after the assessments. It is strictly forbidden to collect the assessment fees from mining enterprises in any form and use assessments for illegitimate interests.

(4) The third-party assessment agencies and experts must keep confidential all the privacy or trade secrets of mining enterprises involved in the assessment work. Data provided by mining enterprises shall be kept and used reasonably and strictly prohibited from being used for other purposes without permission.

2. Assessment requirements

(1) Based on nine industry standards, including the Notice of the General Office of the Ministry of Natural Resources on the Selection of Green Mines in 2020 and the Construction Specifications of Green Mine in the Non-metallic Mine Industry, and relevant policies and technical standards such as the Evaluation Indicators of green mine construction, the assessment shall be carried out in a scientific, rigorous, fair and just way.

(2) The third-party assessment agency shall form an assessment team according to the regulations. The team leader shall be a full-time staff member of the unit, and the team shall have no less than five members. The assessment experts should be familiar with green mines' relevant policies and standards, covering geology, mining, mineral processing, ecological environment, and other related majors.

(3) The third-party assessment report should include the basis, process, summary, and conclusion of the assessment. It is necessary to describe the assessment procedures, consult the data, and check the site conditions to clarify the basis for determining whether the mining enterprises meet the standards' requirements. The sources of evidence documents for the key contents cited should be provided to ensure credible evidence and information, precise content, and accurate judgment. The assessment report and related records/materials (copies) shall be kept in the assessment agency for more than 12 months.

(4) The third-party assessment work should be standardized and rigorous, and the base number of person-days for the onsite verification should not be less than five person-days. Relevant work shall be

subject to the supervision of the competent departments of natural resources, mining enterprises, and the public. The competent departments of natural resources shall circulate the notices of criticism to the third-party assessment agencies verified to have serious violations such as accepting property and divulging the mining enterprise's trade secrets, practicing fraud, and playing favoritism, and committing irregularities. Besides, the assessment reports from the agencies shall not be accepted.

3. Working procedures

(1) Determining the third-parties. After the start of the selection, the provincial competent departments of natural resources shall fill in the proposed list of third-party assessment agencies within their respective administrative areas in the National Green Mines List Management Information System (website: http://greenmine.mnr.gov.cn, hereinafter referred to as the List System), and establish a corresponding relationship with mining enterprises.

(2) Assessment of interior work. The assessment team shall log into the List System, review the electronic version of the self-assessment reports and other data submitted by mining enterprises, and verify the evaluation indicators' prerequisites. When relevant contents and materials are missing, the mining enterprises shall be informed to supplement or resubmit. After a preliminary assessment, the key points of onsite assessment shall be determined through discussion, and the time of onsite assessment shall be informed to mining enterprises.

(3) Onsite assessment. (1) Holding an assessment meeting. The assessment team shall hold an assessment meeting with the relevant management personnel of the mining enterprise, and the leader of the assessment team shall introduce the purpose and plan of the assessment and read out the evaluation work commitment letter. The person in charge of the green mine construction in the mining enterprise shall introduce the construction situation. The assessment team may inquire about the relevant situation. (2) Conducting onsite verification. The assessment team shall go to the production area, office area, living area, and the key sites to check the conditions and standardized degree of the site work item by item according to the evaluation index system. The key sites shall include the sites of key energy-consuming processes and equipment, main pollution control and its equipment, hazardous chemical storage, and tailings reservoir. The assessment team shall also consult the relevant production and operation management records, inquire about the relevant indexes, and take photos of key sites to check records.

(4) Preparation of reports. After the comprehensive consideration of the internal work assessment and the onsite assessment, the assessment team shall discuss and determine the assessment conclusion. The highlights of the mining enterprise's construction work shall be summarized, the losing point items and reasons shall be clarified, and the improvement suggestions shall be proposed. The assessment team shall fill in the *Scoring Table of Green Mine Construction Evaluation Index System* online and prepare the third-party assessment report. Simultaneously, the list of the assessment team's experts, the sign-in form of the assessment meeting, the meeting minutes, the onsite verification records, and the live meeting and onsite photos shall be attached to the report.

(5) Feedback results. The third-party assessment agency shall upload the assessment team commitment letter (seal), the third-party assessment report, and the scanned copy of the List System's expert list signature form. According to the actual needs, the provincial natural resources departments may determine whether to conduct the assessment report's expert argumentation.

Appendix:

Commitment Letter

This assessment team promises that in the third-party assessment of green mine selection, we will be objective, fair and realistic, and carry out the assessment in strict accordance with the green mine assessment policies and standards throughout the process. We will not charge any fees for any reason to ensure the authenticity and effectiveness of the assessment conclusion. If there is any violation or fraud, we are willing to bear the corresponding responsibility.

The third-party agency (Seal):
The leader of the assessment team:
Month/Day/Year

4. Specification of the *Green Mine Evaluation Index*

The *Green Mine Evaluation Index* includes prerequisites and scoring tables. The prerequisites are veto items. If any item fails to meet the requirements, the mining enterprise cannot participate in the green mine selection. Provinces (autonomous regions and municipalities) can add veto items according to laws, regulations, and the actual situation.

1. Scoring method

(1) There are 100 items in the evaluation index scoring table, with a total score of 1,000 points. The green mine construction level shall be scored from six categories: mining environment, resource development mode, comprehensive utilization of resources, energy conservation and emission

reduction, scientific and technological innovation and intelligent mines, and the management and image of enterprises.

(2) Treatment of items not involved. For the score calculation of mining enterprises that do not involve Items 33-36 of the third-level index, it shall be determined whether Items 33-36 belong to the items not involved according to the *Standards for Comprehensive Exploration and Evaluation of Mineral Resources* (GB/T 25283–2010), and it shall be explained in the scoring table. If it belongs to the item not involved, the major category's final score shall be scored by the conversion method. For example, a mine does not involve Items 33-36, and if the total score of Items 37–42 is 64, then the final score in "3. Comprehensive utilization of resources" is "64 / 80 * 120 = 96 points".

2. Compliance specification

(1) In principle, the total score shall not be less than 800 points. The natural resources management departments of all provinces (autonomous regions and municipalities) can appropriately adjust the specific "standard line" according to each region's actual situation without lowering the comprehensive requirements.

(2) In principle, the first-level index (converted score) score shall not be lower than 75% of the index's full score. For example, the full score of the first-level index of "mining area environment" is 220 points, and the actual score of the index shall not be less than 165 points.

3. Scoring specification

(1) If an index is a deduction item in the scoring specification, the index's score shall be deducted until it is finished. If an index is an increased item in the scoring specification, the score will increase until the index's full score increases.

(2) A score must be supported by evidence, and the supporting materials must be kept. The reasons for getting the relevant scores shall be indicated in the column of "verification record." If there is no supporting material or evidence, no score shall be given. If many contents need to be filled in, they can be majorly described in the assessment report.

(3) For the supporting systems such as the centralized construction system of mineral processing, the association relationships should be transparent, and they can be unified into the assessment consideration.

(4) The assessment methods such as questionnaire, onsite examination, and averaging of experts' scores need to be described in the "inspection record" description.

(5) The contents that need to be checked on site should be indicated in the "inspection record", and the contents shall include staff, equipment, facilities, plant, environment, and site.

Table 1 shows the summary of the green mine evaluation index.

Table 1 Summary of green mine evaluation index

First-level index		Second-level index		
Categories	Score	Name	Score	Number of items
1. Mining environment	220	Appearance of mines	170	12
		Plantation in mining area	50	5
2. Resource development mode	240	Resources exploitation	80	2
		Mineral processing and smelting	60	1
		Mine environment restoration and land reclamation	60	4
		Environmental management and monitoring	40	8
3. Comprehensive utilization of resources	120	Comprehensive utilization of associated resources	40	4
		Solid waste disposal and comprehensive utilization	40	3
		Wastewater disposal and comprehensive utilization	40	3
		Comprehensive utilization	40	1
		Solid waste disposal and comprehensive utilization	40	1
		Wastewater disposal and comprehensive utilization	40	2
4. Energy conservation and emission reduction	200	Energy-saving and consumption-reducing	35	4
		Waste gas emission	55	5
		Wastewater emission	50	4
		Solid waste emission	30	1
		Noise emission	30	3
5. Scientific and technological innovation and intelligent mines	115	Scientific and technological innovation	65	8
		Intelligent mines	50	7
6. Management and image of an enterprise	105	Green mine management system	28	5
		Enterprise culture	13	4
		Enterprise management	44	8
		Community harmony	10	2
		Enterprise integrity	10	3
Total	1,000		1,000	100

Note Comprehensive utilization of resources: the first three items of the second-level index are for non-metal, chemical, gold, metallurgy, non-ferrous metal, petroleum, and coal industries, with a total score of 120 points. The last three items are for sand and gravel, cement limestone, and building materials industries, with a total score of 120 points. In the assessment, any one of the above categories shall be selected for different minerals

2 Basic Knowledge

1. ***Law of the People's Republic of China on the Prevention and Control of Environmental Pollution by Solid Waste***

Law of the People's Republic of China on the Prevention and Control of Environmental Pollution by Solid Waste

(Adopted at the 16th meeting of the Standing Committee of the Eighth National People's Congress on October 30, 1995. Revised for the first time at the 13th meeting of the Standing Committee of the Tenth National People's Congress on December 29, 2004. According to the *Decision on Amending Twelve Laws, the First Amendment includes the Cultural Relics Protection Law of the People's Republic of China* at the third meeting of the Standing Committee of the Twelfth National People's Congress on June 29, 2013. According to the *Decision on Amending Seven Laws, the second amendment includes the Port Law of the People's Republic of China* at the 14th meeting of the Standing Committee of the 12th National People's Congress on April 24, 2015. According to the *Decision on Amending Twelve Laws, the third amendment includes the Foreign Trade Law of the People's Republic of China* at the 24th meeting of the Standing Committee of the 12th National People's Congress on November 7, 2016. Revised for the second time at the 17th meeting of the Standing Committee of the 13th National People's Congress on April 29, 2020.)

Catalog

Chapter 1 General Provisions

Article 1 The law is enacted to protect and improve the ecological environment, prevent and control solid waste pollution of the environment, safeguard public health, maintain ecological safety, promote ecological civilization, and sustainable economy and society.

Article 2 This law shall apply to the prevention and control of environmental pollution by solid waste.

This law shall not apply to the prevention and control of marine environmental pollution by solid waste or environmental pollution by radioactive solid waste.

Article 3 The State pursues green development methods to promote cleaner production and a circular economy.

The State advocates a simple and moderate, green and low-carbon lifestyle, and guides the public to prevent and control environmental pollution by solid wastes actively.

Article 4 The prevention and control of environmental pollution by solid wastes adhere to reduction, recycling, and harmlessness.

Any unit and individual shall reduce the amount of solid waste generated, promote its comprehensive utilization, and reduce its hazards.

Article 5 The prevention and control of environmental pollution by solid wastes adhere to responsibility for pollution.

Units and individuals that generate, collect, store, transport, utilize, and treat solid wastes must take measures to prevent or reduce the environmental pollution caused by solid waste, and responsible for the environmental pollution according in accordance with the law.

Article 6 The State pursues a domestic-waste classification system.

Domestic waste classification adheres to government promotion principles, public participation, urban and rural planning, adjusting measures to local conditions, and simple and convenient operation.

Article 7 Local people's governments at various levels shall be responsible for the prevention and control of environmental pollution by solid wastes in their respective administrative regions.

The State implements a target responsibility system and an assessment and evaluation system for the prevention and control of environmental pollution by solid waste, and incorporates the completion of solid waste prevention and control targets into the content of the assessment and evaluation.

Article 8 The people's governments at various levels shall strengthen leadership over the prevention and control of environmental pollution by solid wastes, and organize, coordinate, and urge relevant departments to perform the duties of supervision and management of the prevention and control of environmental pollution by solid waste in accordance with the law.

Provinces, autonomous regions, and municipalities directly under the Central Government can negotiate to establish a joint prevention and control mechanism for environmental pollution caused by solid waste across administrative regions, and coordinate planning and formulation, facility construction, and solid-waste transfer.

Article 9 The competent department of the State Council's ecological environment shall implement unified supervision and management of the prevention and control of environmental pollution by solid waste throughout the country. The State Council's development and reform, industry and information technology, natural resources, housing and urban–rural construction, transportation, agriculture and rural areas, commerce, health, and customs are responsible for the supervision and management of solid waste pollution prevention and control within their respective responsibilities.

The ecological-environment department of the local people's government shall implement unified supervision and management of the prevention and control of environmental pollution by solid wastes in the administrative area. The competent departments of the local people's government for development and reform, industry and information technology, natural resources, housing and urban–rural construction, transportation, agriculture and rural areas, commerce, health, etc. are responsible for the supervision and management of the prevention and control of environmental pollution by solid wastes within their respective responsibilities.

Article 10 The State encourages and supports scientific research, technological development, advanced technology promotion, and scientific popularization of the prevention and control of environmental pollution by solid wastes, and strengthens scientific and technological support for the prevention and control of environmental pollution by solid waste.

Article 11 State agencies, social organizations, enterprises and institutions, grassroots mass autonomous organizations and the news media shall strengthen publicity and education and scientific popularization of the prevention and control of environmental pollution by solid waste, and enhance public awareness of prevention and control of environmental pollution by solid waste.

Schools shall carry out popularization and education of domestic waste classification and the prevention and control of environmental pollution by other solid waste.

Article 12 People's governments at various levels shall commend and reward units and individuals that have achieved outstanding successes in the prevention and control of environmental pollution by solid waste and related extensive use in accordance with relevant national regulations.

Chapter 2 Supervision and Management.

Article 13 People's governments at or above the county level shall incorporate the prevention and control of environmental pollution by solid waste into the national economic and social development and ecological environmental protection plans, and take measures for reducing the amount of solid waste generated, promoting the comprehensive utilization of solid waste, and reducing the harmfulness and the landfill amount of solid waste.

Article 14 The competent department of the ecological environment under the State Council shall, in conjunction with relevant departments under the State

Council, formulate solid wastes identification standards, identification procedures and national technical standards for the prevention and control of environmental pollution by solid waste in accordance with national environmental quality standards and national economic and technological conditions.

Article 15 The standardization authority under the State Council shall, in conjunction with the development and reform, industry and information technology, ecological environment, agriculture and rural areas, and other competent authorities under the State Council, formulate standards for the comprehensive utilization of solid waste.

The comprehensive utilization of solid wastes shall comply with the laws and regulations on the ecological environment and meet the technical standards for preventing and controlling environmental pollution by solid waste. The use of solid waste comprehensive utilization products shall comply with the State's purposes and standards.

Article 16 The competent department of the ecological environment under the State Council shall, in conjunction with relevant departments under the State Council, establish a national information platform for the prevention and control of environmental pollution caused by solid wastes such as hazardous wastes, and promote the monitoring and information traceability of solid waste collection, transfer, and disposal.

Article 17 The construction of the projects that generate, store, utilize, and dispose of solid wastes, environmental impact assessments shall be carried out according to the law, and observe the national regulations on construction projects' environmental protection management.

Article 18 The supporting facilities for the prevention and control of environmental pollution by solid wastes determined in the environmental impact assessment documents of a construction project shall be designed, constructed, and used simultaneously with the main project. The preliminary design of a construction project shall, in accordance with the requirements of environmental protection design specifications, incorporate the prevention and control of environmental pollution by solid waste into the environmental impact assessment document, implement measures to prevent and control environmental pollution by solid waste and damage the ecology, and investment estimates for solid waste pollution prevention and control facilities.

In accordance with the provisions of relevant laws and regulations, the construction unit shall inspect and accept the supporting facilities to prevent and control environmental pollution by solid waste and prepare an acceptance report to open to the public.

Article 19 Units and other producers and operators that collect, store, transport, utilize, and dispose of solid waste shall strengthen the management and maintenance of relevant facilities, equipment, and sites to ensure their normal operation and use.

Article 20 Units and other producers and operators that produce, collect, store, transport, utilize and treat solid waste shall take measures to prevent the spread, loss, leakage, or environmental pollution, and shall not dump, stack, discard or scatter solid wastes without authorization.

Any unit or individual is prohibited from dumping, stacking, and storing solid waste to rivers, lakes, canals, channels, reservoirs, beach lands and bank slopes below their highest water level, and other locations specified by laws and regulations.

Article 21 It is forbidden to construct facilities and sites for the centralized storage, utilization, and disposal of industrial solid waste and hazardous waste, and domestic garbage landfills in the red line areas of ecological protection, concentrated areas of permanent basic farmland, and other areas that need special protection.

Article 22 If solid waste is transferred out of the administrative area of a province, autonomous region, or municipality directly under the Central Government for storage and disposal, an application shall be submitted to the competent department of ecological environment of the people's government of the province, autonomous region, or municipality where the solid waste is removed. The competent department of ecological environment of the people's government of the province, autonomous region, and municipality directly under the Central Government of the place of transfer shall promptly negotiate with the competent department of ecological environment of the people's government of the receiving province, autonomous region, or municipality directly under the Central Government, and approve the transfer of the solid waste out of the province, autonomous region, or municipality within the prescribed time limit. Without approval, no transfer is allowed.

Suppose solid wastes are transferred out of the administrative area of a province, autonomous region, or municipality directly under the Central Government. In that case, it shall be reported to the competent department of ecological environment of the people's government of the province, autonomous region, or municipality where the solid waste is removed for record. The competent department of ecological environment of the people's government of the province, autonomous region, or municipality directly under the Central Government of the place of transfer shall notify the competent department of ecological environment of the people's government of the province, autonomous region, or municipality directly under the Central Government of the receiving place of the filing information.

Article 23 Solid waste outside the people's Republic of China shall be prohibited from entering the country for dumping, stacking, and disposal.

Article 24 The State gradually realizes zero import of solid waste, which shall be organized and implemented by the State Council's competent department of ecology and environment in conjunction with the State Council's competent departments of commerce, development and reform, and customs.

Article 25 The customs may entrust a professional organization to identify the imported goods suspected of being solid wastes, and manage them according to the identification conclusion.

Article 26 The competent department of ecological environment and the environmental law enforcement agencies and other departments responsible for the supervision and management of the prevention and control of environmental pollution by solid waste shall, within the scope of their respective functions and duties, have the right to conduct on-site inspections on the units involved in wastes and other activities and producers and operators engaged in producing, collecting, storing, transporting, utilizing and disposing solid waste. The persons under inspection shall provide the necessary information to report the situation.

The implementation of on-site inspections may take measures such as on-site monitoring, collecting samples, consulting, or duplicating materials related to the prevention and control of environmental pollution by solid wastes. Inspectors shall present their certificates when conducting on-site inspections. The business secrets learned during the on-site inspection shall be kept confidential.

Article 27 In one of the following circumstances, the competent department of the ecological environment and other departments responsible for the supervision and management of the prevention and control of environmental pollution by solid waste may deal with the illegal collection, storage, transportation, utilization, and disposal of solid wastes and facilities, equipment, premises, tools, and articles shall be sealed up and seized:

(1) May cause evidence to be lost, concealed, or illegally transferred;
(2) Causing or likely to cause severe environmental pollution.

Article 28 The competent department of the ecological environment shall, in conjunction with relevant departments, establish a credit record system for entities and other producers and operators that generate, collect, store, transport, utilize, and dispose of solid waste, and incorporate relevant credit records into the national credit information sharing platform.

Article 29 The competent department of ecological environment of the people's government of a city divided into districts shall, in conjunction with the competent departments of housing and urban–rural construction, agriculture and rural areas, and health, regularly release to the public the types, output, disposal capacity, utilization, and disposal status of solid wastes.

Units discharging, collecting, storing, transporting, using, or treating hazardous waste shall promptly disclose information on the prevention and control of environmental pollution by solid wastes and actively accept social supervision.

Units using and treating hazardous waste shall open facilities and places to the public in accordance with the law to increase public awareness and participation in environmental protection.

Article 30 The people's governments at or above the county level shall incorporate the prevention and control of environmental pollution by industrial solid wastes, domestic wastes, hazardous wastes and other solid wastes into the annual report on the environmental status and the completion of environmental protection goals, and report to the people's Congress or the Standing Committee at the corresponding level.

Article 31 Any units and individuals have the right to report the units and individuals that pollute the environment by solid wastes.

The competent department of ecological environment and other departments responsible for the supervision and management of the prevention and control of environmental pollution by solid waste shall publish the reporting methods on preventing and controlling environmental pollution by solid wastes, thus facilitating public reporting.

The department that receives the report shall promptly deal with it and keep the reporter's relevant information confidential. For the real-name report that verified the truth, rewards will be given.

Suppose the whistleblower reports to the unit to which he/she belongs. In that case, the unit shall not retaliate against the whistleblower by dissolving or changing the labor contract or other means.

Chapter 3 Industrial Solid Wastes.

Article 32 The competent department of the ecological environment under the State Council shall, in conjunction with the competent departments of development and reform, industry and information technology under the State Council, define the hazards and impacts of industrial solid wastes on public health and the ecological environment, formulate technical policies for the prevention and control of environmental pollution by industrial solid wastes, and promote advanced production process and equipment for the prevention and control of environmental pollution by industrial solid wastes.

Article 33 The competent department of industry and information technology under the State Council shall, in conjunction with relevant departments under the State Council, study and promote production techniques and equipment that reduce industrial solid wastes and the hazards of industrial solid wastes, and publish a list of backward production techniques and equipment producing industrial solid wastes within a time limit.

Producers, sellers, importers, and users shall stop the production, sale, import, or use of the equipment listed in the catalog stipulated in the preceding paragraph within the time limit prescribed by the competent department of industry and information technology in conjunction with relevant departments under the State Council. The adopter of the production techniques shall stop using the technology listed in the list specified in the preceding paragraph within the time limit prescribed by the competent department of industry and information technology in conjunction with the relevant departments under the State Council.

The eliminated equipment included in the list of elimination within a time limit shall not be transferred to others for use.

Article 34 The competent department of industry and information technology under the State Council shall, in conjunction with the competent departments of development and reform and the ecological environment under the State Council, regularly publish an oriented catalog of technologies, techniques, equipment, and products for the comprehensive utilization of industrial solid wastes, organize the evaluation of comprehensive utilization of industrial solid waste resources, and promote the comprehensive utilization of industrial solid waste.

Article 35 Local people's governments at or above the county level shall formulate plans for the prevention and control of environmental pollution by industrial solid waste, organize the construction of facilities for the centralized disposal of industrial solid waste, and promote the prevention and control of environmental pollution by industrial solid wastes.

Article 36 Units that produce industrial solid wastes shall establish a sound environmental pollution prevention and control system for the entire process of industrial solid waste generation, collection, storage, transportation, utilization, and disposal, establish an industrial solid waste management account to accurately record the production of industrial solid wastes by truthfully recording information on the type, quantity, flow direction, storage, utilization, and disposal of wastes, thus enabling industrial solid waste to be traceable and inquired, and taking measures to prevent and control industrial solid wastes from polluting the environment.

It is forbidden to throw industrial solid waste into the collection facilities of domestic wastes.

Article 37 If a unit that generates industrial solid wastes entrusts others to transport, utilize, or dispose industrial solid wastes, the unit shall verify the principal qualifications and technical capabilities of the entrusted party, and sign a written contract in accordance with the law to stipulate pollution prevention requirements in the contract.

The entrusted party who transports, utilizes, and disposes of industrial solid wastes shall fulfill the requirements for pollution prevention and control in accordance with the provisions of relevant laws and regulations and contractual agreements, and inform the units that produce industrial solid waste of the transportation, utilization, and disposal.

Suppose a unit producing industrial solid wastes violates the provisions of the first paragraph of this article, in addition to being punished in accordance with the relevant laws and regulations. In that case, it shall bear joint and liability with the entrusted party causing environmental pollution and ecological damage.

Article 38 Units that produce industrial solid wastes shall implement clean production audits in accordance with the law, rationally select and utilize raw materials, energy, and other resources, adopt advanced production techniques and equipment, and reduce industrial solid wastes and its harmfulness.

Article 39 Units that produce industrial solid waste shall obtain a pollutant discharge permit. The specific measures and implementation steps for pollutant discharge permits shall be prescribed by the State Council.

Units producing industrial solid waste shall provide relevant information, such as the type, quantity, flow direction, storage, utilization, and disposal of industrial solid waste, to the local ecological and environmental authorities, and specific measures to reduce the generation of industrial solid waste and promote comprehensive utilization, and implement the relevant regulations of the pollution-discharge-permit management system.

Article 40 Units that produce industrial solid wastes shall utilize industrial solid wastes in accordance with economic and technical conditions. Suppose solid wastes are temporarily unused or cannot be used. In that case, storage facilities and places shall be constructed following the regulations of the State Council's ecological environment and other competent authorities, with safe classification and store or harmless disposal measures. The storage of industrial solid waste shall adopt protective measures that meet the national environmental protection standards.

The construction of facilities and sites for industrial-solid-waste storage and disposal shall meet the national environmental protection standards.

Article 41 If a unit producing industrial solid waste terminates, it shall take pollution prevention and control measures at the facilities and places where industrial solid wastes are stored and disposed, and properly dispose of undisposed industrial solid wastes to prevent environmental pollution.

In case of any change in the unit producing industrial solid wastes, the changed unit shall, in accordance with the provisions of the State on environmental protection, safely dispose the untreated industrial solid waste and its storage and disposal facilities and sites, or take adequate measures to ensure the safe operation of such facilities and sites. If the parties concerned have agreed otherwise on the responsibilities for preventing and controlling industrial solid waste pollution and the facilities and sites for their storage and disposal prior to the change, such agreement shall prevail. However, the parties shall not be exempted from their obligations of pollution prevention and control.

The cost of safe disposal of undisposed industrial solid wastes and storage and disposal facilities and sites that have been terminated before April 1, 2005 shall be borne by the relevant people's government. However, if the unit's land use right enjoyed is transferred according to law, the transferee of the land use right shall bear the disposal costs. If the parties agree otherwise, the agreement shall prevail; however, the parties shall not be exempted from their pollution prevention obligations.

Article 42 Mining enterprises shall adopt scientific mining methods and beneficiation techniques to reduce the production and storage of mining solid wastes such as tailings, coal gangue, and waste rock.

The State encourages the comprehensive utilization of mining solid wastes such as tailings, coal gangue, and waste rock by adopting advanced technologies.

After the storage facilities for mining solid wastes such as tailings, coal gangue, and waste rock are not used, mining enterprises shall close the sites in accordance with relevant national environmental protection regulations to prevent environmental pollution and ecological damage.

Chapter 4 Domestic Wastes.

Article 43 Local people's governments at or above the county level shall speed up establishing a domestic waste management system of classified release, collection, transportation, and treatment to cover the domestic wastes classification system.

Local people's governments at or above the county level shall establish a coordination mechanism for domestic waste classification and strengthen and coordinate the classification and management of domestic wastes.

The people's governments at various levels and the relevant departments shall organize and carry out publicity on domestic waste classification to educate and guide the public to develop the habit of classification of domestic wastes and urge and guide domestic wastes classification.

Article 44 Local people's governments at or above the county level shall systematically improve the fuel structure, develop clean energy, and reduce solid wastes such as fuel wastes.

The local people's government's relevant departments at or above the county level shall strengthen the management of the production and circulation of products, avoid excessive packaging, organize the listing of clean vegetables, and reduce domestic wastes generated.

Article 45 The people's government at or above the county level shall make overall arrangements for the construction of urban and rural domestic garbage collection, transportation, and treatment facilities, determine the site of the facilities, improve the comprehensive utilization and harmless disposal of domestic wastes, and industrialization of domestic wastes collection and treatment to establish and improve a social service system for the prevention and control of environmental pollution by domestic wastes.

The local people's government's relevant departments at or above the county level shall make overall planning, rationally arrange recycling, sorting, and packing outlets to promote the recycling utilization of domestic wastes.

Article 46 Local people's governments at all levels shall strengthen the prevention and control of environmental pollution by rural domestic wastes, and protect and improve the rural human settlement environment.

The State encourages the source reduction of rural domestic wastes. In urban–rural junctions, densely populated rural areas, and other places where conditions permit, an integrated urban–rural household waste management system should be established. Other rural areas should actively explore domestic waste management, adopt measures to local conditions, use the wastes nearby, or properly dispose domestic waste.

Article 47 The competent environmental sanitation department of the people's government at or above the city level with districts shall formulate standards for the construction and operation of facilities and sites for cleaning, collection, storage, transportation, and treatment of domestic waste, publish a guidance catalogue for domestic waste classification to strengthen supervision and management.

Article 48 The environmental sanitation and other competent departments of the local people's government at or above the county level shall organize the cleaning, collection, transportation, and treatment of urban and rural domestic garbage, and select qualified units to engage in the cleaning, collection, transportation and processing of domestic wastes through bidding, etc.

Article 49 Units, households, and individuals generating domestic waste shall perform the obligation of reducing the source of domestic wastes and sorting out in accordance with the law, and take the responsibility of the producer of domestic waste.

Any units and individuals shall sort and put domestic wastes in designated places in accordance with the law. It is forbidden to dump, throw, pile, or burn domestic garbage at will.

Organs, institutions, etc. should play a leading role in the classification of domestic waste.

Domestic garbage that has been sorted and put in should be collected, transported, and disposed in accordance with regulations.

Article 50 The cleaning, collection, transportation, and treatment of urban and rural domestic wastes shall abide by the relevant state regulations on environmental protection and environmental sanitation management to prevent environmental pollution.

If hazardous wastes classified and collected from domestic wastes are classified as harmful ones, they shall be managed in accordance with harmful wastes.

Article 51 Business units engaged in public transportation shall promptly clean up and collect domestic waste generated during transportation.

Article 52 Farmer's markets, agricultural product wholesale markets, etc., shall strengthen environmental sanitation management, and maintain environmental clean, with the timely sweeping, separate collection, and proper disposal of generated garbage.

Article 53 Units engaged in the construction and development of new urban areas, old urban areas, residential quarters, and villages and towns, as well as the operation and management units of public facilities and places such as airports, wharves, stations, parks, shopping malls, stadiums, and other public facilities and places, shall construct supporting facilities for the collection of domestic wastes in accordance with the relevant provisions of the State on environmental sanitation.

The local people's government at or above the county level shall coordinate the effective connection of public transport and treatment facilities for domestic wastes with the collection facilities specified in the preceding paragraph, and integrate the integration of the domestic wastes classification collection and transportation system and the renewable resource recovery system in planning, construction, and operation.

Article 54 The materials recovered from domestic waste shall be used in accordance with the purposes and standards prescribed by the State, and shall not be used for the products that may endanger human health.

Article 55 The construction of domestic waste-treatment facilities and sites shall comply with the environmental protection and sanitation standards stipulated by the competent departments of the ecological environment and housing and urban–rural development under the State Council.

The construction of domestic waste-treatment facilities in neighboring areas should be encouraged to promote the co-construction and sharing across administrative regions.

It is forbidden to close, leave idle, or dismantle domestic wastes treatment facilities and sites without authorization. Suppose it is necessary to be closed, left idle, or dismantled. In that case, it shall be approved by the competent environmental sanitation department of the local municipal or county-level people's government and the local ecological environment competent department to prevent environmental pollution.

Article 56 The domestic waste-treatment units shall install and use monitoring equipment according to relevant national regulations, monitor pollutants' discharge and promptly disclose the pollution discharge data. The monitoring equipment shall be networked with the monitoring equipment of the local ecological environment authority.

Article 57 The local people's government's environmental sanitation department at or above the county level shall organize and carry out the recycling and harmless treatment of kitchen wastes.

Units and other producers and operators producing and collecting kitchen wastes shall hand over the kitchen wastes to units with corresponding qualifications for harmless treatment.

It is forbidden for livestock farms and breeding communities to use kitchen wastes that have not been harmlessly treated to feed livestock and poultry.

According to the generator pays principle, Article 58 Local people's governments shall establish a domestic waste-disposal fee system at or above the county level.

According to local conditions concerning the classification of domestic wastes, local people's government at or above the county level shall formulate the charging standards for domestic waste treatment, embodying differentiated management such as classified pricing, metering and charging, and soliciting public opinions. The charging measures for domestic waste-disposal shall be announced to the public.

The domestic waste-treatment fee shall be used exclusively for their collection, transportation, and treatments, and not for other purposes.

Article 59 Provinces, autonomous regions, municipalities directly under the Central Government, cities divided into districts, and autonomous prefectures may formulate specific local domestic wastes management methods based on actual conditions.

Chapter 5 Construction Wastes and Agricultural Solid Wastes.

Article 60 Local people's governments at or above the county level shall strengthen the prevention and control of environmental pollution caused by construction wastes by establishing a classification and treatment system.

Local people's governments at or above the county level shall formulate work plans to prevent and control environmental pollution caused by construction waste, including source reduction, classified treatment, disposal facilities, site layout, and construction.

Article 61 The State encourages adopting advanced technologies, techniques, equipment, and management measures to reduce construction wastes at the source and establish a construction-waste recycling system.

Local people's governments at or above the county level shall promote the comprehensive utilization product application of construction wastes.

Article 62 The environmental sanitation department of the local people's government at or above the county level shall prevent and control environmental pollution caused by construction wastes, establish a whole-process management system, regulate the generation, collection, storage, transportation, utilization, and disposal, and promote comprehensive utilization, and the construction of waste disposal facilities and sites to ensure the safety of disposal and prevent environmental pollution.

Article 63 The project construction units shall formulate a construction wastes treatment plan, adopt pollution prevention and control measures, and report to the local people's government's environmental sanitation authority at or above the county level for the record.

The project construction units shall remove and transport solid wastes such as construction wastes generated during the construction of the project, and use or dispose them in accordance with the regulations of the environmental sanitation authority.

The project construction units shall not dump, throw or stack the construction wastes generated during engineering construction without authorization.

Article 64 The agricultural and rural administrative departments of the people's governments at or above the county level shall guide the establishment of agricultural solid wastes recycling systems, encourage and guide relevant units and other producers and operators to collect, store, transport, utilize, and dispose agricultural solid wastes in accordance with the law, and strengthen supervision manage to prevent environmental pollution.

Article 65 Units and other producers and operators producing agricultural solid wastes, such as straw, discarded agricultural film, pesticide packaging wastes, shall take recycling and other measures to prevent environmental pollution.

Those engaged in large-scale livestock and poultry breeding shall collect, store, utilize, or dispose solid wastes such as livestock and poultry manure generated promptly to avoid environmental pollution during the breeding process.

It is prohibited to burn straw in the open in densely populated areas, around airports, near main traffic lines, and other areas designated by the local people's government.

The State encourages research and development, production, sales, and use of agricultural films degradable and harmless in the environment.

Article 66 The State establishes an extended producer responsibility system for electrical and electronic products, lead-storage batteries, and automotive power batteries.

Producers of electrical and electronic products, lead storage batteries, and vehicle power batteries shall establish a waste product recycling system that matches product sales by self-built or commissioned methods according to regulations and publish it to achieve recycling and utilization.

The State encourages product producers to develop ecological design to promote resource recycling.

Article 67 The State implements multi-channel recycling and centralized disposal system for waste electrical and electronic products.

It is forbidden to hand over abandoned motor vehicles and ships to enterprises or individuals that do not meet the prescribed conditions for recycling or dismantling.

The dismantle, utilization, and disposal of waste electrical and electronic products, waste motor vehicles, and ships shall comply with relevant laws and regulations and take measures to prevent environmental pollution.

Article 68 The design and manufacture of products and packaging shall comply with the State's cleaner production regulations. The department in charge of standardization under the State Council shall organize the formulation of relevant standards in accordance with national economic and technical conditions, the prevention and control of environmental pollution by solid wastes, and the technical requirements of products to prevent excessive packaging from environmental pollution.

Producers and operators should abide by mandatory standards that restrict the excessive packaging of goods and avoid excessive packaging. The market supervision and management departments and relevant departments of the local people's governments at or above the county level shall strengthen the supervision and management of excessive packaging in accordance with their respective duties.

According to the law, enterprises producing, selling, and importing products and packaging materials included in the mandatory recycling catalog shall recover the products and packaging materials according to relevant State regulations.

E-commerce, express delivery, and take-out industries should prioritize reusable and easily recyclable packages-optimizing, reducing and recycling the packaging. The commerce and postal departments of the local people's government shall strengthen supervision and management at or above the county level.

The State encourages and guides consumers to use green packaging and reduced-quantity packaging.

Article 69 The State prohibits and restricts the production, sale, and use of non-degradable plastic bags and other disposable plastic products in accordance with the law.

Units, e-commerce platform companies, express companies, and take-out companies in commodity-retailing sites shall report the use and recycling of plastic bags and other disposable plastic products to the competent departments of commerce and post in accordance with relevant national regulations.

The State encourages and guides disposable plastic products such as plastic bags to be reduced and recycled actively, and promotes alternative products that are renewable, easily recyclable, and degradable alternative products.

According to relevant national regulations, Article 70 The tourism and accommodation industries shall promote non-proactive provision of disposable items.

Offices of government agencies, enterprises and institutions shall use products, equipment and facilities conducive to environmental protection rather than disposable office supplies.

Article 71 The maintenance and operation units of urban sewage treatment facilities or the sludge treatment units shall safely treat the sludge, ensure that the treated sludge meets the relevant national standards, track and record the flow direction, use, and consumption of the sludge, and report to the urban drainage authority and the ecological environment authority.

The competent urban drainage departments of the people's governments at or above the county level shall incorporate the sludge treatment facilities into the urban drainage and sewage treatment planning, promote the simultaneous construction of sludge and sewage treatment facilities, and encourage collaborative treatment. The collection standard and compensation scope of sewage treatment fees shall cover the cost of sewage sludge treatment and the normal operating cost of sewage treatment facilities.

Article 72 It is prohibited to dump, stack, or discard sludge produced by urban sewage-treatment facilities and treated sludges without authorization.

Sludges with heavy metals or other toxic and hazardous substances exceeding standard are forbidden to enter agricultural land.

The dredging of water bodies shall deal with the bottom sludges produced during dredging in accordance with relevant national regulations to prevent environmental pollution.

Article 73 Various laboratories at all levels and their establishment units shall strengthen the solid wastes' management generated in the laboratory by collecting, storing, transporting, using, and disposing in accordance with the law. If laboratory solid wastes are hazardous, they shall be managed following hazardous waste.

Chapter 6 Hazardous Wastes.

Article 74 The provisions of this chapter shall apply to preventing and controlling environmental pollution by hazardous wastes. If there are no provisions in this chapter, other relevant provisions of this law shall apply.

Article 75 The competent department of the ecological environment under the State Council shall, in conjunction with relevant departments under the State Council, formulate a national hazardous waste list to stipulate uniform hazardous waste identification standards, methods, marks, and identification units' management requirements. The national hazardous waste list should be adjusted dynamically.

The competent department of the ecological environment under the State Council scientifically evaluates the environmental risks of hazardous wastes based on the hazardous characteristics and the amount, implements hierarchical and classified management, establishes an informationized supervision system, and manages and shares hazardous waste transfer data and information through informationized means.

Article 76 The people's governments of provinces, autonomous regions, and municipalities directly under the Central Government shall organize relevant departments to formulate plans for the construction of hazardous waste centralized disposal facilities and sites, scientifically evaluate hazardous-waste disposal, arrange hazardous-waste centralized disposal facilities and sites, and ensure the disposal of hazardous wastes in their respective administrative areas.

The establishment of construction planning for facilities and sites for the centralized disposal of hazardous wastes shall solicit the opinions of relevant industry associations, enterprises and institutions, experts, and the public.

Neighboring provinces, autonomous regions, and municipalities directly under the Central Government can develop regional cooperation to coordinate the construction of regional hazardous waste centralized disposal facilities and sites.

Article 77 For hazardous waste containers and packaging, and facilities and places for the collection, storage, transportation, utilization, and disposal of hazardous wastes, identification signs shall be set up in accordance with regulations.

Article 78 Units producing hazardous wastes shall formulate management plans in accordance with relevant national regulations. The establishment of hazardous wastes management accounts shall truthfully record relevant information and declare the hazards types, amount, flow direction, storage, and disposal to the local ecological and environmental authorities through the national hazardous waste information management system.

The hazardous waste management plan referred to in the preceding paragraph shall include reducing hazardous wastes and their hazards, and storing, utilizing, and disposing hazardous wastes. The hazardous-waste management plan shall be reported to the competent department of the ecological environment where the units producing hazardous wastes for the record.

If units that produce hazardous wastes have obtained a pollutant discharge permit, the pollutant-discharge-permit management system's provisions shall be implemented.

Article 79 Units producing hazardous wastes shall store, utilize, and dispose hazardous wastes in accordance with relevant national regulations and environmental protection standards, rather than dump or stack without authorization.

Article 80 Units engaged in the business activities of collecting, storing, utilizing, and disposing hazardous wastes shall apply for a permit in accordance with relevant state regulations. The specific management measures for permits shall be formulated by the State Council.

It is forbidden to engage in business activities such as the collection, storage, utilization, and disposal of hazardous wastes without a license or in accordance with the provisions of the license.

It is prohibited to provide or entrust hazardous waste to unlicensed units or other producers and operators to collect, store, utilize, and dispose activities.

Article 81 The collection and storage of hazardous wastes shall be carried out according to hazardous-waste characteristics. It is forbidden to collect, store, transport, and dispose of hazardous wastes that are incompatible in nature without safely disposal.

The storage of hazardous waste shall take protective measures that meet the national environmental protection standards. It is forbidden to mix hazardous wastes with non-hazardous wastes for storage.

Units engaged in the business activities of collecting, storing, utilizing and disposing hazardous wastes shall not store hazardous wastes for more than one year. If it is necessary to extend the period, the units shall report to the competent department of the ecological environment that issued the permit for approval; except as otherwise provided by laws and administrative regulations.

Article 82 During transferring hazardous wastes, the electronic or paper transfer form shall be filled out and operated in accordance with relevant State regulations.

For the transfer of hazardous waste across provinces, autonomous regions, and municipalities directly under the Central Government, an application shall be made to the competent department of ecological environment of the people's government of the province, autonomous region, or municipality where the hazardous waste are transferred. The competent department of ecological environment of the people's Government of the province, autonomous region or municipality directly under the central government of the place where the hazardous wastes are to be removed shall, after obtaining the consent of the competent Ecological Environment Department of the people's Government of the receiving Province, autonomous region or municipality directly under the central government, approve the transfer of the hazardous wastes within the prescribed time limit, and notify the competent ecological environment department and Transportation Department of the people's Government of the relevant province, autonomous region or municipality directly under the central government. No transfer should be made without approval.

Hazardous-waste-transfer management should be managed and controlled throughout the process to increase efficiency. The specific measures shall be formulated by the competent department of the ecological environment in conjunction with the competent department of transportation and public security under the State Council.

Article 83 In the transportation of hazardous wastes, measures shall be taken to prevent environmental pollution and comply with the national regulations on dangerous-goods transportation management.

It is prohibited to carry hazardous wastes and passengers on the same means of transportation.

Article 84 When the sites, facilities, equipment, containers, packaging, and other articles for collecting, storing, transporting, utilizing, and disposing hazardous wastes are used for other purposes, they shall undergo decontamination treatment in accordance with relevant national regulations before use.

Article 85 Units producing, collecting, storing, transporting, using, or disposing hazardous wastes shall formulate preventive measures and emergency plans for accidents in accordance with the law, and report to the local ecological and environmental authorities and other departments responsible for supervision and administration of preventing and controlling environmental pollution caused by solid wastes for the record. The competent department of ecological environment and other departments responsible for supervision and administration of preventing and supervising environmental pollution by solid wastes shall conduct inspections.

Article 86 Units that have caused severe environmental pollution by hazardous wastes due to accidents or other emergencies shall immediately take measures to eliminate or reduce the pollution hazards to the environment, promptly inform the units and residents that may be harmed by the pollution, and report to the local ecological and environmental authorities and relevant departments to accept investigation and settlement.

Article 87 In the event or there is evidence that hazardous wastes may occur and seriously pollute the environment and threaten the safety of residents' lives and property, the competent department of ecological environment or other departments responsible for the supervision and management of preventing and controlling environmental pollution by solid wastes shall immediately report to the government and the relevant departments of the people's government at a higher level. The people's government shall take effective measures to prevent or reduce harm. The relevant people's government may order the suspension of operations that cause or may cause environmental pollution accidents as needed.

Article 88 Prior to the decommissioning of key hazardous wastes centralized disposal facilities and sites, the operating units shall adopt pollution prevention and control measures in accordance with relevant State regulations. Expenses for decommissioning shall be pre-provisioned, included in the investment budget or production cost, exclusively for that of key hazardous wastes centralized disposal facilities and sites. The specific extraction and management shall be formulated by the financial department and the price authority in conjunction with the State Council's ecological environment authority.

Article 89 It is forbidden to transfer hazardous wastes via the People's Republic of China's territory.

Article 90 Medical wastes shall be managed in accordance with the national hazardous wastes list. Local people's governments at or above the county level shall strengthen the capacity building for the centralized disposal of medical wastes.

The competent departments of health and ecological environment of the people's governments at or above the county level shall strengthen the supervision and management of collecting, storing, transporting, and treating medical wastes within their respective duties to prevent harm to public health and environmental pollution.

According to law, medical and health institutions shall classify and collect their units' medical wastes and hand them over to the centralized medical wastes disposal units for disposal. The centralized disposal units of medical wastes shall collect, transport and dispose medical wastes promptly.

Medical and health institutions and medical wastes centralized disposal units shall prevent the loss, leakage, infiltration, and spread of medical wastes.

Article 91 When emergencies such as major infectious disease epidemics occur, the people's governments at or above the county level shall coordinate the collection, storage, transportation, and disposal of medical and other hazardous wastes to ensure the necessary vehicles, venues, disposal facilities, and protective materials. The competent departments of health, ecological environment, environmental sanitation, and transportation shall coordinate and cooperate to perform emergency response duties in accordance with the law.

Chapter 7 Safeguard Measures.

Article 92 The relevant departments of the State Council, local people's governments at or above the county level and their relevant departments shall when compiling territorial and spatial and related special plans, coordinate the facilities construction demand for transferring and centralized treating solid wastes such as domestic wastes, construction wastes, and hazardous wastes to guarantee the land for transfer, centralized disposal facilities.

Article 93 The State adopts economic and technical policies and measures conducive to preventing and controlling environmental pollution by solid wastes, supports relevant parties to adopt related measures, and strengthens the training and guidance of personnel engaged in preventing and controlling environmental pollution by solid wastes to promote the professional and large-scale development of solid-waste pollution prevention and control industry.

Article 94 The State encourages and supports scientific research units, solid wastes generation units, solid wastes utilization units, and solid wastes disposal units to tackle key problems jointly, and develop new technologies for the comprehensive utilization and centralized disposal of solid wastes, as well as the skills of preventing and controlling environmental pollution by solid wastes.

Article 95 People's governments at all levels shall strengthen the prevention and control of environmental pollution by solid wastes and arrange necessary funds for the following matters in accordance with the principle of powers division:

(1) Scientific research and technological development for preventing and controlling environmental pollution by solid wastes.
(2) Classification of domestic wastes.
(3) Construction of solid-waste centralized disposal facilities.
(4) Emergency treatment of medical and other hazardous wastes arising from emergencies such as major infectious diseases.
(5) Other matters related to prevent and control environmental pollution by solid wastes.

The use of funds should strengthen performance management and audit supervision to ensure effectiveness.

Article 96 The State encourages and supports social forces to prevent and control environmental pollution by solid wastes, and provides policy support in accordance with relevant State regulations.

Article 97 The State develops green finance to encourage financial institutions to increase credit for solid wastes pollution prevention projects.

Article 98 Those engaged in preventing and controlling environmental pollution by solid wastes such as the comprehensive utilization shall enjoy tax preferences in accordance with the provisions of laws and administrative regulations.

The State encourages and advocates all sectors of society to donate property for preventing and controlling environmental pollution by solid wastes with preferential taxation in accordance with laws and administrative regulations.

Article 99 Units collecting, storing, transporting, using, and treating hazardous wastes shall, in accordance with the relevant provisions of the State, take out environmental pollution liability insurance.

Article 100 The State encourages units and individuals to purchase and use comprehensive utilization and reusable products.

People's governments at or above the county level and their relevant departments shall, in government procurement, prioritize purchasing comprehensive utilization and reusable products.

Chapter 8 Legal Liability.

Article 101 The competent department of ecological environment or other departments responsible for supervision and administration of preventing and controlling environmental pollution by solid wastes violates the provisions of this law and commits one of the following acts, the people's government at the same level or the relevant department of that at a higher level shall instruct to correct, and punish the directly responsible persons in charge and other directly responsible persons according to the law:

(1) Failing to issue administrative licenses or approval documents according to the law.
(2) Covering up illegal acts.

(3) Failing to seal up or detain according to the law.
(4) Failing to investigate illegal acts after discovering or receiving reports.
(5) Committing other illegal acts such as abuse of power, dereliction of duty, and malpractice for personal gain.

Where an administrative penalty decision should have been made in accordance with the provisions of this Law, the higher-level competent department may directly make an administrative penalty.

Article 102 Violation of this law's provisions and any of the following behaviors shall be ordered to be corrected by the competent department of ecological environment, fined, and illegal gains confiscated. If the circumstances are severe, it shall be reported to the people's government that has the right to approve and ordered to suspend business or close down:

(1) Units generating, collecting, storing, transporting, using, or treating solid wastes fail to disclose information on preventing and controlling environmental pollution by solid wastes promptly according to the law.
(2) The domestic wastes treatment units failing to install and use monitoring equipment in accordance with relevant national regulations, monitor the discharge of pollutants in real-time, and disclose pollution discharge data.
(3) Transferring the eliminated equipment included in the elimination list within a time limit to others for use.
(4) Constructing facilities and sites for centralized storing, using, and treating industrial solid and hazardous wastes, and domestic wastes landfills, in areas with ecological protection red lines, permanent basic farmland concentrated areas and other areas that need special protection.
(5) Transferring solid wastes out of the administrative area of a province, autonomous region, or municipality directly under the Central Government for storage and disposal without approval.
(6) Transferring solid wastes out of a province's administrative area, autonomous region, or municipality directly under the Central Government without reporting for the record.
(7) Dumping, stacking, discarding, or throwing industrial solid wastes without authorization, or failing to take corresponding preventive measures, causing industrial solid wastes to spread, lose, leak, or other environmental pollution.

    Units producing industrial solid wastes having not established a management account with truthful record.
(8) Units producing industrial solid wastes entrust others to transport, utilize, or dispose industrial solid wastes in violation of this law's provisions.
(9) The storage of industrial solid wastes failing to take protective measures that meet the national environmental protection standards.

(10) Units and other producers and operators violating other solid-waste management requirements, polluting the environment, and destroying the ecology.

Anyone who commits one of the first and eighth acts of the preceding paragraph shall be fined 50 up to 200 thousand yuan. Anyone who commits one of the second to eleventh items of the preceding paragraph shall be fined 100 thousand up to one million yuan. Anyone who commits the seventh act of the preceding paragraph shall be fined one to three times the required disposal cost. If the required disposal cost is less than 100 thousand yuan, the fine shall be calculated as 100 thousand yuan. If the relevant laws and administrative regulations provide otherwise for the penalties for the acts in item eleventh of the preceding paragraph, those provisions shall apply.

Article 103 Anyone who violates the provisions of this law refuses or obstructs supervision and inspection by delaying, confining, or detaining law enforcement personnel, or resorts to fraud in accepting supervision and inspection, the departments responsible for supervision and administration of preventing and controlling environmental pollution caused by solid wastes shall order corrections with a fine from 50 to 200 thousand yuan. The directly responsible persons in charge and other directly responsible persons shall be fined from 20 to 100 thousand yuan.

Article 104 Anyone who violates the provisions of this law and produces industrial solid wastes without the pollutant discharge permit in accordance with the law shall be ordered by the competent department of ecological environment to make corrections or restrict and stop production for remediation with a fine from 100 thousand to 1 million yuan. If the circumstances are severe, it shall be reported to the people's government that has the right to approve and ordered to suspend business or close down.

Article 105 Producers and operators violating the provisions of this law without complying with mandatory standards restricting excessive packaging shall be ordered by the local people's government's market supervision and management department at or above the county level or relevant departments to make corrections. Those who refuse to make corrections shall be fined from 2 to 20 thousand yuan. If the circumstances are serious, a fine from 20 to 100 thousand yuan shall be imposed.

Article 106 Anyone violating the provisions of this law without complying with national regulations on prohibiting or restricting the use of non-degradable plastic bags and other disposable plastic products, or reporting the use of plastic bags and other disposable plastic products in accordance with relevant national regulations, the commerce and postal departments of the local people's government at or above the county level shall order corrections with a fine from 10 to 100 thousand yuan.

Article 107 Anyone who engages in large-scale livestock and poultry farming and fails to collect, store, utilize or dispose solid wastes such as livestock and poultry manure generated during the breeding shall be ordered to make corrections by the competent department of the ecological environment with a fine less than 100 thousand yuan. If the circumstances are serious, it shall report to the people's government that has the right to approve and ordered to suspend business or close down.

Article 108 In violation of this law's provisions, the maintenance and operation units of urban sewage treatment facilities or the sludge treatment units fails to track and record the flow direction, use, and consumption of sludge, or the treated sludge does not meet the relevant national standards, the urban drainage authority shall order corrections with warnings. If serious consequences are caused, a fine from 100 to 200 thousand yuan shall be imposed. If they refuse to make corrections, the urban drainage department may designate a unit with the governance capacity to govern, with the offender bearing the cost.

Anyone who violates the provisions of this law, dumps, stacks, discards or leaves the urban drainage department shall order the sludge produced by urban sewage treatment facilities and treated sludge without authorization to make corrections with a fine from 200 thousand to 2 million yuan, and the directly responsible persons in charge and other directly responsible persons shall be fined from 20 to 100 thousand yuan. If serious consequences are caused, the directly responsible persons in charge and other directly responsible persons shall be fined from 50 to 500 thousand yuan. If they refuse to make corrections, the urban drainage department may designate a unit with the governance capacity to govern, with the offender bearing the cost.

Article 109 Anyone violating the provisions of this law by producing, selling, importing, or using obsolete equipment, or adopting obsolete production techniques, shall be ordered by the department designated by the local people's government at or above the county level to make corrections with a fine from 100 thousand to one million yuan and confiscate illegal gains. If the circumstances are serious, the department designated by the local people's government at or above the county level shall put forward opinions to report to the authorized people's government for approval and order it to suspend business or close down.

Article 110 After the mining solid wastes storage facilities such as tailings, coal gangue, waste rock, etc., are not closed in accordance with the relevant national environmental protection regulations, the competent department of the ecological environment shall order corrections with a fine from 200 thousand to one million yuan.

Article 111 If one of the following acts is committed in violation of this law's provisions, the environmental sanitation department of the local people's government at or above the county level shall order it to make corrections, impose a fine, and confiscate the illegal gains:

(1) Anyone dumping, throwing, stacking or burning domestic wastes at will.
(2) Anyone closing, leaving idle, or dismantling domestic wastes treatment facilities or sites without authorization.
(3) Construction units failing to prepare a construction wastes treatment plan and report it for the record or to remove the solid wastes generated during the construction promptly.
(4) Construction units dumping, throwing or stacking construction wastes generated during the construction without authorization, or failing to use or dispose the solid wastes generated during the construction in accordance with regulations.
(5) Units and other producers and operators producing and collecting kitchen wastes failing to hand over the kitchen wastes to units with corresponding qualifications for harmless treatment.
(6) Livestock and poultry farms and breeding communities using kitchen wastes without harmlessly treated to feed livestock and poultry.
(7) Anyone discarding domestic waste along the way during transportation.

Units having one of the first and seventh acts of the preceding paragraph shall be fined from 50 to 500 thousand yuan. Units having one of the second to sixth acts of the preceding paragraph shall be fined from 10 thousand to 1 million yuan. Individuals who commit one of acts in Items 1, 5 and 7 shall be fined from 1 to 5 hundred yuan.

Anyone who violates this law's provisions and fails to separate household wastes in designated locations shall be ordered by the environmental sanitation department of the local people's government at or above the county level to make corrections. If the circumstances are severe, units shall be fined from 50 to 500 thousand yuan, and individuals shall be fined according to the law.

Article 112 Anyone violating this law's provisions and any of the following behaviors shall be ordered to be corrected by the competent department of ecological environment, fined, and illegal gains confiscated. If the circumstances are severe, it shall be reported to the people's government that has the right to approve and ordered to suspend business or close down:

(1) Anyone failing to set up hazardous wastes identification signs in accordance with regulations.
(2) Anyone failing to formulate hazardous wastes management plans or declare hazardous wastes related materials in accordance with relevant State regulations.
(3) Anyone dumping or stacking hazardous wastes without authorization.
(4) Anyone providing or entrusting hazardous wastes to unlicensed units or other producers and operators to engage in business activities.

(5) Anyone failing to fill in the hazardous wastes transfer form in accordance with relevant State regulations or transferring hazardous wastes without approval.
(6) Anyone failing to store, utilize, or dispose hazardous wastes in accordance with national environmental protection standards or mix hazardous wastes with non-hazardous wastes for storage.
(7) Anyone indistinguishably collecting, storing, transporting, and treating hazardous wastes of incompatible properties without safe disposal.
(8) Anyone carrying hazardous wastes and passengers on the same means of transportation.
(9) Anyone using sites, facilities, equipment, containers, packaging and other items for collecting, storing, transporting, and disposing hazardous wastes for other purposes without decontamination treatment.
(10) Anyone failing to take corresponding preventive measures to cause the dispersion, loss, leakage, or other environmental pollution of hazardous wastes.
(11) Anyone discarding or throwing away hazardous wastes along the way during transportation.
(12) Anyone failing to formulate preventive measures and emergency plans for hazardous wastes accidents.
(13) Anyone failing to establish and truthfully record hazardous wastes management accounts in accordance with relevant state regulations.

Anyone committing one of the acts in Items 1, 2, 5, 6, 7, 8, 9, 12, and 13 of the preceding paragraph shall be fined from 100 thousand to one million yuan. Anyone committing one of the acts in Items 3, 4, 10, and 11 of the preceding paragraph shall be fined from three to five times the required disposal cost. If the necessary disposal cost is less than 200 thousand yuan, the fine shall be calculated as 200 thousand yuan.

Article 113 Producers of hazardous wastes who violate the provisions of this law and fail to dispose the hazardous wastes produced in accordance with the regulations and refuse to make corrections after being ordered, the competent department of ecological environment shall organize the disposal on its behalf, and the producers shall pay the expenses. Anyone refusing to bear the cost of handling on behalf of him/her shall be fined from one to three times the cost.

Article 114 Anyone engaged in the business activities of collecting, storing, utilizing, and disposing hazardous wastes without a permit shall be ordered by the competent department of ecological environment to make corrections, and fined from 1 to 5 million yuan, and be reported to the people's government with the right to approve and ordered to suspend business or close down. The legal representative, the main responsible person, the directly responsible person in charge, and other responsible persons shall be fined from 100 thousand to one million yuan.

Anyone failing to engage in the business activities of collecting, storing, utilizing, and disposing hazardous wastes in accordance with the provisions of the license shall be ordered by the competent department of ecological environment to make corrections, restrict and stop production for rectification, and impose a fine from 500 thousand to 2 million yuan. The legal representative, the main responsible person, the directly responsible person in charge, and other responsible persons shall be fined from 50 to 500 thousand yuan. If the circumstances are serious, it shall be reported to the people's government with the right to approve and ordered to suspend business or close down, and the license issuing authority may also revoke the license.

Article 115 Anyone who violates this law's provisions and transfers hazardous wastes through the People's Republic of China shall be ordered by the customs to return the solid wastes with a fine from 500 thousand to 5 million yuan.

The carrier shall be jointly and severally liable with the importer for the return and disposal of the solid wastes specified in the preceding paragraph.

Article 116 Anyone who violates the provisions of this law and transfers hazardous wastes through the People's Republic of China shall be ordered by the customs to return the hazardous wastes and impose a fine of 500 thousand up to 5 million yuan.

Article 117 For solid wastes that have entered the country illegally, the competent department of ecological environment of the people's government at or above the provincial level shall submit its disposal opinions to the customs in accordance with the law, and the customs shall make a penalty decision in accordance with Article 115 of this law. If environmental pollution has been caused, the competent department of the people's government's ecological environment at or above the provincial level shall order the importer to eliminate the pollution.

Article 118 Anyone who violates the provisions of this law and causes an environmental pollution accident caused by solid waste shall, in addition to assuming the liability for compensation in accordance with the law, be imposed by the competent department of ecological environment in accordance with the provisions of paragraph two of this article and ordered to take control measures within a time limit. Those who cause serious or severe environmental pollution accidents by solid wastes may also be reported to the people's government with the right to approve and order to close down.

Suppose a general or relatively large accident of environmental pollution caused by solid wastes happens. In that case, a fine shall be calculated based on one to three times of the accident's direct economic losses. For a major or very serious accident of environmental pollution caused by solid wastes, a fine shall be calculated based on three to five times of the direct economic losses of the accident, and the legal representative, main responsible person, directly responsible person in charge and other responsible persons shall be imposed a fine of less than 50% of the annual income from the unit.

Article 119 Units and other production and business operators discharging solid wastes in violating of this law's provisions should be fined and ordered to make corrections, and the administrative agency making the penalty decision according to the law shall organize a recheck. Those committing such illegal acts shall be punished consecutively daily in accordance with the provisions of the Environmental Protection Law of the People's Republic of China.

Article 120 If one of the following acts in violation of the provisions of this law does not constitute a crime, the public security organ shall detain the legal representative, the main person in charge, the directly responsible person in charge, and other responsible persons for a period from 10 to 15 days. Those with less serious circumstances shall be detained for 5 to 10 days:

(1) Anyone dumping, stacking, discarding, or dropping solid wastes without authorization, causing serious consequences.
(2) Anyone constructing facilities, sites and domestic wastes landfills for centralized storing, using, and treating industrial solid and hazardous wastes in areas with ecological protection red lines, permanent basic farmland concentrated areas, and other areas that need special protection.
(3) Anyone providing or entrusting hazardous wastes to unlicensed units or other producers and operators for stacking, utilization, or disposal.
(4) Anyone engaging in the business activities of collecting, storing, utilizing, and disposing hazardous wastes without a license or in accordance with the provisions of the license.
(5) Anyone transferring hazardous waste without approval.
(6) Anyone failing to take preventive measures to cause the dispersion, loss, leakage, or other serious consequences of hazardous wastes.

Article 121 Where solid wastes pollute the environment, destroy the ecology or harm national and social public interests, relevant agencies and organizations may file a lawsuit in the people's court in accordance with the Environmental Protection Law of the People's Republic of China, the Civil Procedure Law of the People's Republic of China, the Administrative Procedure Law of the People's Republic of China and other laws.

Article 122 Where solid wastes pollute the environment or damage the ecology and cause severe losses to the country, the local people's government at or above the level of a city divided into districts or its designated institutions and organizations shall negotiate with the organizations and other producers and operators that cause environmental pollutions and ecological damages to require them to bear the liability for damages. If the negotiation fails to reach an agreement, a lawsuit may be filed in the people's court.

The local people's government at or above the county level shall organize to disposal solid wastes that cannot be identified or returned in the course of law enforcement.

Article 123 If a violation of this law constitutes a violation of public security management, the public security organ shall impose public security management penalties in accordance with the law. If a crime is constituted, criminal responsibility shall be investigated in accordance with the law. If personal or property damage is caused, civil liability shall be borne in accordance with the law.

Chapter 9 Supplementary Provisions

Article 124 The meanings of the following terms used in this Law:

(1) Solid wastes refer to solid, semi-solid, or gaseous articles and substances in containers that have lost their original use values or have been discarded or given up although having not lost their utilization values, as well as the articles and substances included in the management of solid wastes stipulated by laws and administrative regulations, unless they are harmlessly processed, meet the mandatory national product quality standards, and will not endanger public health or ecological safety, or they are not classified as solid wastes according to the solid waste identification standards and procedures.
(2) Industrial solid wastes means solid wastes discharged from production activities of industries.
(3) Domestic wastes mean solid wastes generated in daily life or activities providing services for daily life, and these regarded as domestic wastes according to laws and administrative regulations.
(4) Construction wastes refer to the abandoned soil, materials, and other solid wastes generated by the construction unit during building, rebuilding, expanding, and dismantling buildings, structures, and pipe networks, as well as the residents' decoration.
(5) Agricultural solid waste means solid waste discharged from production activities of agriculture.
(6) Hazardous wastes mean solid wastes with hazardous characteristics included in the national hazardous waste list or identified in accordance with the hazardous waste identification standards and methods prescribed by the State.
(7) Storage refers to the activity of temporarily placing solid wastes in specific facilities or places.
(8) Utilization means the activity of extracting substances from solid wastes as raw materials or fuels.
(9) Disposal means burning solid wastes and changing the physical, chemical, and biological characteristics of solid wastes to reduce the solid

wastes, reduce or eliminate its dangerous components, or place the solid wastes into the landfill sites meeting environmental protection requirements.

Article 125 This law shall apply to the prevention and control of pollution of liquid wastes. However, the prevention and control of pollution by wastewater discharged into water bodies shall be governed by other relevant laws, instead of this law.

Article 126 This law shall enter into force as of September 1, 2020.

2. Open-Pit Mining

Open-pit mining should achieve dust-free operations in the entire production process (perforation, blasting, mining and loading, transportation, windrow, and slag discharge), and the discharge, vibration, and noise pollution of the three wastes must meet national standards. Therefore, it will not form the sites that difficult for geological, environmental management, and land reclamation. In the mining process, unified blasting, professional blasting, medium-length blasting, bench mining, and mechanized shovel loading operation are implemented. Meanwhile, self-blasting, floor cut blasting, expanding springing blasting, and one-sided wall mining operations are prohibited.

The open-pit mines' mining technology includes perforation, blasting, mining and loading, transportation, and rock discharge.

2.1 Open-pit mining process

(1) Perforation

Perforation is the primary process of open-pit mining, accounting for 10 to 15% of the total production cost.

A. Down-the-hole drill

Down-the-hole drill varies widely in drilling angles with high mechanization, which improves the operation rate and reduces auxiliary operation time. Besides, its flexibility with low weight and investment costs enable it to control the ore′s grade by drilling inclined holes, which improves blasting quality and eliminates roots and bulk. Therefore, the down-the-hole drills, suitable for perforating medium-hard ore, are adopted in small and medium-sized mines at home and abroad.

B. Rotary blasthole drill

As new drilling equipment developed from the rotary drill, the rotary blasthole drill has the characteristics of high efficiency, mechanization and automation, and low cost. Thus, it can be used to perforate ore-rock operations with different hardness, making perforation equipment commonly used in open-pit mines worldwide.

C. Rock drilling jumbo

Rock drilling jumbo is new rock drilling equipment emerging in the mining industry. It installs one or more rock drills and automatic propellers together on a special drill boom or bench, which mechanizes the rock drill's operation with a walking mechanism.

(2) Blasting

Blasting is to crush the hard solid ore to provide suitable excavated materials for the mining and loading work, accounting for 15 to 20% of the open-pit mining's total cost. The blasting quality affects the mining and loading efficiency, transportation, coarse crushing, and the mine's total cost.

A. Shallow hole blasting

The blasthole diameter of shallow hole blasting is 30 to 75 mm, while the blasthole depth is below 5 m, sometimes up to 8 m. If drilling with a rock drilling jumbo, the blasthole depth can be increased.

Shallow hole blasting is used to deal with small-scale open-pit sand and gravel mines or quarries, chambers, tunnel excavation, secondary blasting, and new open-pit mine bag, forming transportation pathways open-pit single-wall ditch on hillsides, and other blasting.

B. Deep hole blasting

Deep hole blasting is a method that uses drilling equipment to drill a deeper hole as a charged space for mining explosives. The deep-hole blasting in open-pit mines is dominated by bench production blasting.

The drilling equipment of deep hole blasting can be used for down-the-hole drills and rotary blasthole drills to drill deep vertical holes and inclined blastholes. The charging of the inclined blastholes is uniform, with the excellent blasting quality of the ore-rock, which can create the right conditions for mining and loading work.

Under certain conditions, large-area micro-delay blasting, intermediate charges in the blastholes, or bottom air gap charges can be adopted to reduce the seismic effect and blasting costs, improve the blasting quality, and achieve economic benefits.

C. Chamber blasting

Chamber blasting is a method of blasting large explosives in the tunnel of the blasting chamber. Open-pit mines are used under capital construction and specific conditions, while quarries are used with available conditions and high mining demand.

D. Multi-row-hole millisecond blasting

With the increase of excavator bucket capacity and open-pit mine production capacity, the blasting required for regular mining and blasting in open-pit mining increases. Therefore, large-scale blasting such as multi-row-hole millisecond blasting and multi-row-hole millisecond extruding blasting are used in open-pit mining at home and abroad.

The advantages of multi-row-hole millisecond blasting are:

1) The extensive blasting reduces the blasting frequency and time to avoid shots, which improves the utilization of stope equipment.
2) The crushing quality of ore-rock is improved, and the boulder frequency is 40 to 50% less than that of single-row-hole blasting.
3) The increase of the working time′s utilization coefficient and the decrease of the operation times of perforating equipment and post-blasting filling areas improve the efficiency of perforating equipment by 10 to 15%.
4) The efficiency of mining, loading, and transportation equipment increases by 10 to 15%.

E. Multi-row-hole millisecond extruding blasting

Multi-row-hole millisecond extruding blasting means the multi-row-hole millisecond blasting with the blast pile remains on the working face. Muck bank is conducive to extrusion: Firstly, the effective time′s extension of blasting can improve explosives' utilization and crushing effects. Then, the width control of the muck piles can prevent the ore from flying away. Besides, the millisecond interval of multi-row-hole millisecond extruding blasting is 30 to 50% larger than that of ordinary blasting, while 50 to 100 ms is preferred in China.

The advantages of multi-row-hole millisecond extruding blasting are:

1) The crushing effect of ore-rock is better. The front muck bank's obstruction makes the first and subsequent rows of drillholes increase the charge, broking under the muck bank's extrusion.
2) Muck piles are concentrated. Mines transported by rail cannot be demolished before blasting, improving the equipment's efficiency of mining, loading, and transportation.

The disadvantages of multi-row-hole millisecond extruding blasting are as follows:

1) The explosive consumption is large.
2) The working platform should be broader enough to accommodate the muck bank.
3) The height of muck piles affects the operation safety of the excavator. Regardless of the blasting method adopted, the blasting operations should implement the *Blasting Safety Regulations* (GB 6722–2011) and set up safety warning signs to ensure personnel and property safety with good alert work.

F. Blasting measures near the slope

The open-pit mine's downward extension leads to slope stability, so the blasting near the slope should be controlled. Millisecond blasting, pre-splitting blasting, and smooth blasting are adopted worldwide.

Millisecond blasting reduces vibration. The increase of blasting segments and the control of the millisecond interval have a cushioning effect.

Pre-splitting blasting isolates the slope. The steps of pre-split blasting near the slope are as follows: drill a row of dense parallel boreholes along the slope boundary and load them with explosives. Then, detonate before blasting the dass to obtain a crack that penetrating each borehole with a certain width. The pre-crack separates the dass and slope, while the reflection of the seismic waves from the mining blasting on the crack surface is weakened, which protects the slope.

Smooth blasting protects the slope. The smooth blasting steps near the slope are as follows: Drill a row of dense parallel boreholes along the boundary line and load them with explosives. Then, blast after blasting the mining boreholes to form parallel palisades along the dense boreholes. The initiation times of smooth and pre-splitting blastings are different: The initiation of smooth blastholes is 50–75 ms later than that of the mining boreholes in previous rows.

Besides, the blasting of the holes in the last few rows should be controlled. The reduction of the charge and resistance lines of boreholes near the slope in the last few rows reduces the drills blasting damage to the slope, called "buffer blasting".

(3) Mining and loading, and transportation

A. Mining and loading

Mining and loading operations are to excavate ore-rock from underground or muck piles with loaders into the transport machinery's carriage, or the designated place. As the key to open-pit mining, it is served by other production processes, such as blasting and transportation.

Mining and loading equipment includes excavators, draglines, hydraulic shovel, and tire-type front loader.

B. Transportation

In open-pit mining, the investment in the capital construction of mine transportation accounts for 60% of the mine′s, while transportation and labor costs account for more than half of the total ore cost and total labor, respectively, showing the importance of transportation.

Transportation in open-pit mining includes dump truck transportation, railway locomotive transportation, belt conveyor transportation, slope skip hoisting transportation, and combined transportation, where dump truck transportation is the most common (See Table 2).

Mining and loading, and transportation are inseparable in mutual influence and mutual restriction. The development of mining, loading, and transportation technology is reflected in the large-scale mining and transportation equipment, the integration and continuity between mining and loading, and transportation, and computer automation.

(4) Rock dumping

Rock dumping is the transportation terminal's operation, which transports the stripped topsoil and waste rock to the waste-rock yard for disposal.

Rock dumping technology includes railway transportation, highway transportation, and belt transportation.

Table 2 Transportation methods of open-pit mining

Transportation methods	Features and applications
Dump truck transportation	Flexible maneuverability; convenient transportation; strong climbing ability; the maximum slope reaching 10 to 15%
Railway locomotive transportation	Large transportation volume; low cost; Large basic investment; poor flexibility; suitable for open-pit mines with large reserves and area, and long distance
Belt conveyor transportation	Strong climbing ability; high automation; ability of continuous and semi-continuous operation, widely used in open-pit mines

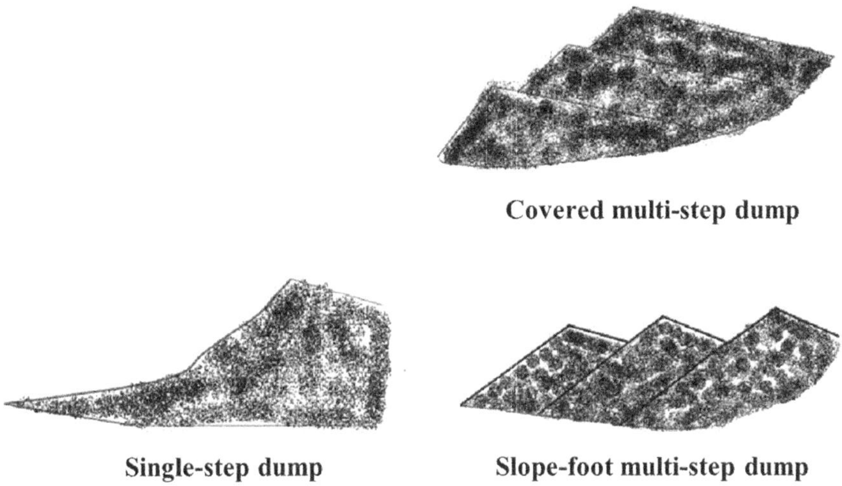

Fig. 1 Different stacking methods of dumps

(5) Dump

Dump (waste-rock yard): A place where stripped materials are stacked, and mining wastes are discharged.

Dumps are divided into the single-step dump, covered multi-step dump, and slope-foot multi-step dump based on the stacking order. Figure 1 shows the stacking methods.

Dumps are divided into road transportation dump, railway transportation dump, belt transportation dump, and hydraulic transportation dump according to the dumping technology.

2.2 Determination of step height in open-pit mines

Step height is a technical and economic indicator of open-pit mining, affecting the mine's economic benefits and the efficiency of the open-pit mining's production technology such as perforation, mining, and loading. The steps with considerable height decrease, which reduces the cost. However, the determination of open-pit mines′ step height must consider economic, technical, and safety factors because of the open-air slope's low stability.

(1) The step height of flat truck loading. The flat truck loading means that the transportation equipment and the excavator work are at the same level. For safety, during digging without pre-cracking loose soft rock, the step height is not higher than the excavator's maximum digging height. During digging a blast pile of hard ore, the step height should make the blast pile's height after blasting not higher than the excavator's maximum digging height. Finally, only when the soft ore's step height and the hard ore's blast pile height are not less than 2/3 of the height of the excavator's pushing shaft could the full-bucket of the excavator increase.

(2) The step height of upper truck loading. The upper truck loading means the transportation equipment locates on the upper flat plate of the step where the excavator is located. The step height should be determined by the maximum unloading height and radius of the excavator to load the ore-rock into the upper flat plate of transportation equipment.

3. **Mining circular economy**

Circular economy (abbreviation for material circulation and mobility economy) is a new economic form and development model. As an economic development model harmonious with the environment, it is characterized by resource conservation and recycles. Figure 2 shows the operation mode of the circular economy.

The circular economy's operation mode regards economic activities as a repeated circulation process of "resources-products-wastes-renewable resources." The differences between this economic growth model and the traditional economy are as follows: its goal is to achieve sustainable development with the principle of coordinating the relationship between human and nature, the core of the efficient and

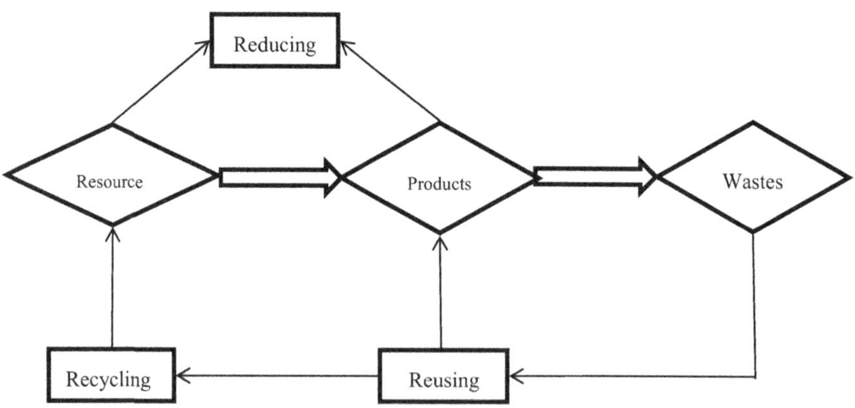

Fig. 2 Operation mode of the circular economy

Table 3 Comparison between circular and traditional economies

Compared projects	Traditional economy	Circular economy
Movement form	The developmental linear economy with the one-way material flow (resource consumption → product industry → pollution discharge)	Network economy of material energy cycle in enclosed type (resource utilization → green industry → resource regeneration)
Resource utilization	Extensive operation with one-time utilization; high mining with low utilization	Resource recycling with scientific operation and management; low mining with high utilization
Waste discharge and its impact on the environment	High waste discharge; cost externalization; Environmentally unfriendly	Zero or low waste discharge; Environmentally friendly
Ambition	Economic benefits (Product profit maximization)	Economic and environmental benefits and society achieve sustainable development
Economic growth mode	Quantitative growth	Intensive development
Environmental governance mode	Terminal treatment	Prevention first with full process control
Supporting theories	Political economy, welfare economics, and other traditional economic theories	Ecosystem and industrial ecology theories (development, governance, and restoration together)
Evaluating indicators	Single economic indicators (GDP, per capita consumption, etc.)	Green accounting system (green GDP, etc.)

recycling utilization of resources, the principle of "reducing, reusing and recycling," and the basic characteristics of low consumption, low emissions, and high efficiency. Table 3 shows the comparison between circular and traditional economies.

3.1 Connotation and characteristics of the circular mining economy

The circular mining economy follows the characteristics and ecological laws of mineral products. The economic system in harmony with the global environment and social progress forms by internally superimposing the closed-loop material flow (consisting of exploration, mining, processing and smelting production, deep processing, and consumption), the dependent energy flow, and the information flow.

(1) The core of the circular mining economy is the comprehensive utilization of mineral resources.
(2) The circular mining economy is different from the traditional mining economy.
(3) The circular mining economy is a feedback process of "mineral exploration, mineral resources, products, renewable mineral resources, and final emissions."

3.2 Understanding of "reducing, Reusing and Recycling"

"Reducing, reusing and recycling" includes three levels:

(1) The product's green design runs through the concept of "reducing, reusing and recycling". The scope of green design includes the following aspects: any design based on understanding for the earth's ecosystem and human living environment; any design that is conducive to the sustainable development of society; any design conducive to the healthy development of human and biological living environments.

(2) Material resources run through the concept of "reducing, reusing and recycling" throughout the development and utilization cycle: rational development and multi-level reuse of resources in the development stage; product recycling and reusing in the design stage of product and production technology; multi-level utilization of resources and the integration and standardization of productive technology in the production technology's system design; the process integration and refuse reclamation in the production process, transportation, and sales; the extension of products' service life and multiple utilization of resources in the stage of circulation and consumption; the reuse of resources, and the recovery and recycling of wastes at the end of the life cycle.

(3) Redevelopment and recycling of eco-environmental resources include reproducing and reusing renewable resources in the environment and the repairing, reuse, and recycling of space and environmental resources.

3.3 Basic mode and implementation method of mining development circular economy

(1) Enterprise level of the circular economy—the circular economy model within enterprises

The enterprise-level of the industrial system in a circular economy belongs to the small cycles, an industry system of internal circular economy in a single enterprise with the microcirculation of materials and energy as the main body. Mining enterprises promote cleaner production to develop a circular economy in the following aspects:

1) Energy-saving and consumption reduction in the production process.
2) Improving material efficiency by utilizing recyclable raw materials.
3) Reducing waste discharge.
4) Improving comprehensive utilization and recycling of resources.
5) Advocating materials circulation between enterprises (zero discharge).

(2) Regional level of circular economy-development areas of green mining.

The circular economy's regional level in the circular economy belongs to the medium cycle, a circular economy for symbiotic enterprises. Industrial metabolism and symbiosis between enterprises are formed by integrating material, energy, and information between enterprises, establishing eco-industrial parks.

The development areas of green mining (ecological mining park) are industrial paragenetic assemblage based on the mining industry. The waste, energy, and by-products that cannot be digested in mining companies are turned into raw materials or power for other enterprises. The two-two combination forms an enterprise network to realize the harmonious flow of logistics, energy, and information flows.

(3) Circular society
The circular economy society is a new momentum in developing the international circular economy, and Germany and Japan are typical. The social circular economy system is manifested in the following aspects:

1) Improving the legislation of circular economy.
2) Improving the economic policy of the circular economy.
3) Carrying out the circular enterprise economy.

3.4 Micro-operation rules for the development of the circular economy in the mining industry

(1) 3R principles in mining activities
3R principle refers to reducing, reusing, and recycling. The manifestation in the mining circular economy is the combination of rich and poor mines, comprehensive recovery, transforming one mine into multiple mines, clean production, and ecological protection.
The general requirements of the 3R principle for the mining production process are embodied in mineral resource utilization and environmental protection. Table 4 shows the 3R principle of circular mining economy.

(2) Exploitation of mineral resources under the reduction principle

1) Reduce the output rate of waste rock.
Measures to reduce the output rate of waste rock from the source are as follows:
Strengthen the production exploration with improved exploration accuracy to avoid over-excavating of waste rocks caused by ineffective exploitation, accurate mining, and cutting.
Design, exploiting, mining and cutting, and arrange them in the vein; open-pit mines should reduce the stripping ratio through research.
Choose a mining method with a high recovery rate and low dilution rate.
Optimize the blasting parameters and technology, especially the ultra-deep blastholes, avoiding over-exploitation and under-exploitation to reduce large lump and fine ore output rates.
After each stope, the goaf is measured to provide an accurate stoping boundary for adjacent stopes' design.
2) Reduce the output rate of tailings.
3) Reduce wastewater discharge.
4) Reduce exhaust emissions.

(3) Exploitation of mineral resources under the recycling principle

Table 4 3R principles of the circular mining economy

Mining production process 3R principle	Reducing principle	Reusing principle	Recycling principle
Exploration	Reduce the amount, extent, and depth of exploration	Improve the quality and accuracy of exploration with advanced theories and instruments	Publish survey data and multiple developments
Mining	Reduce mining and vegetation damage, and mine the rich and poor minerals	Reduce stripping ratio, dilution rate, and increasing mining rate with advanced mining technology; appropriate mining	Reduce the cut-off grade of mineral products with the improvement of processing technology
Mineral Processing	Reduce ores to be selected	Adopt advanced processes, pay attention to comprehensive utilization, moderate the scale of mineral processing, and increase mineral processing recovery rate	Improving the technology and comprehensively recovering useful elements; using tailings, such as backfilling and construction; wastewater utilization
Smelt	Reduce the smelting amount	Use advanced technology to improve the recovery rate of main elements and product quality, save energy, and reduce consumption to maintain a moderate scale	Recycle useful refractory elements and difficult to recycle, comprehensively using waste gas, water, and residue
Deep-processing	Develop new products and new application; waste recycling and reusing	Improving adaptability to achieve standardization, comprehensively recovering useful elements	Harmless disposal of waste

A. Solid wastes recycling

The recycling ways of solid wastes from mines are as follows:

1) Recover useful metals.
2) Recover useful minerals.
3) Recover energy.
4) Prepare materials.

Figure 3 shows the recycling mode of solid wastes from mines.

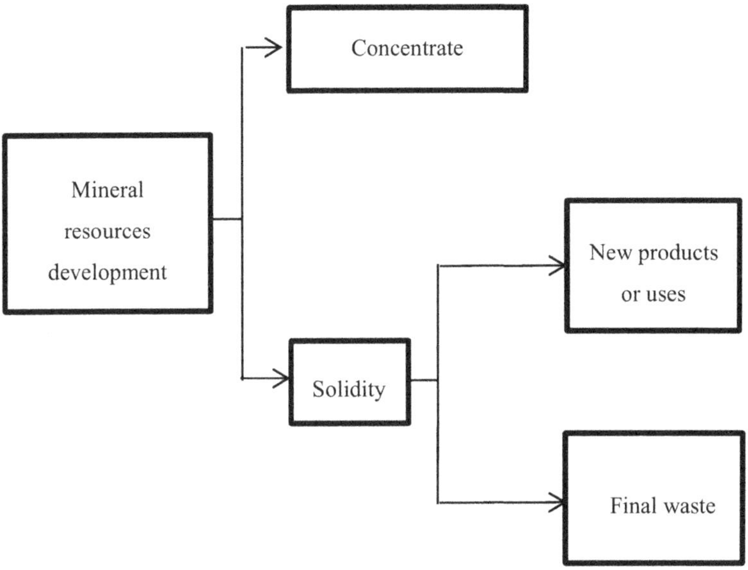

Fig. 3 Recycling model of solid wastes from mines

Figure 4 shows the technical recycling routes of solid wastes from mines.

B. Wastewater recycling

The development of the water recycling economy through establishing a water resource circular economy model in mining areas (See Fig. 5) can solve wastewater pollution and recycle water resources.

C. Waste gas recycling

D. Waste heat and pressure recovering and recycling

In the long run, the optimal use of mineral resources should follow:

(1) Durable use.
(2) Appropriate scale and intensive use.

3.5 Technical support system

Mining circular economy is a technical system of "low mining, high utilization, low emission" formed using ecological laws to guide mining development.

The technology carrier of mining development for a circular economy is environmentally sound technology, which has the following characteristics: low emission of pollution and waste; utilization of resources, energy, and various useful components of ore; the recovery of useful components in wastes, and the disposal of residual and temporarily hard-to-use wastes in an environmentally acceptable way.

(1) Cleaner production technology.
(2) Waste utilization technology.
(3) Comprehensive utilization technology of mineral resources.
(4) Value-added technology of mineral products.

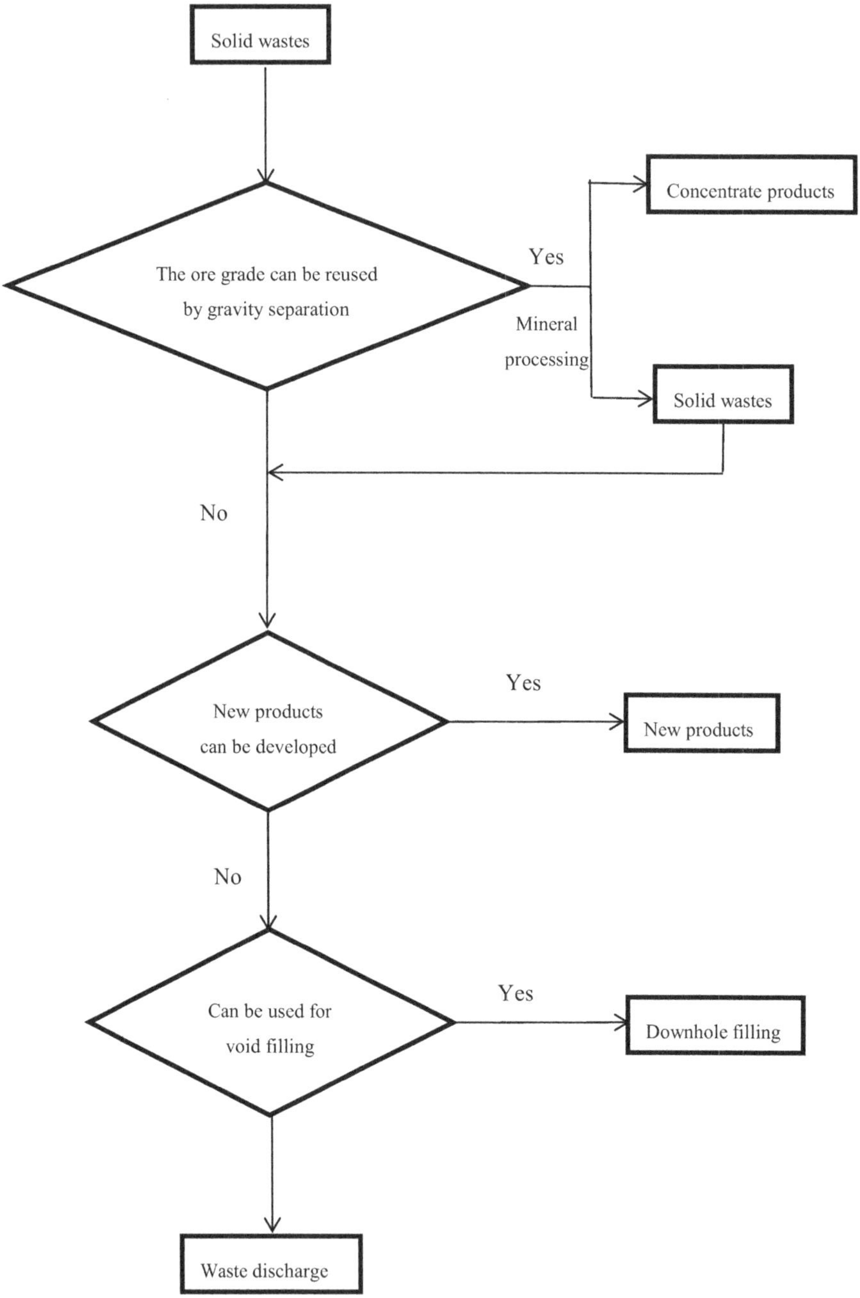

Fig. 4 Recycling technical routes of solid wastes from mines

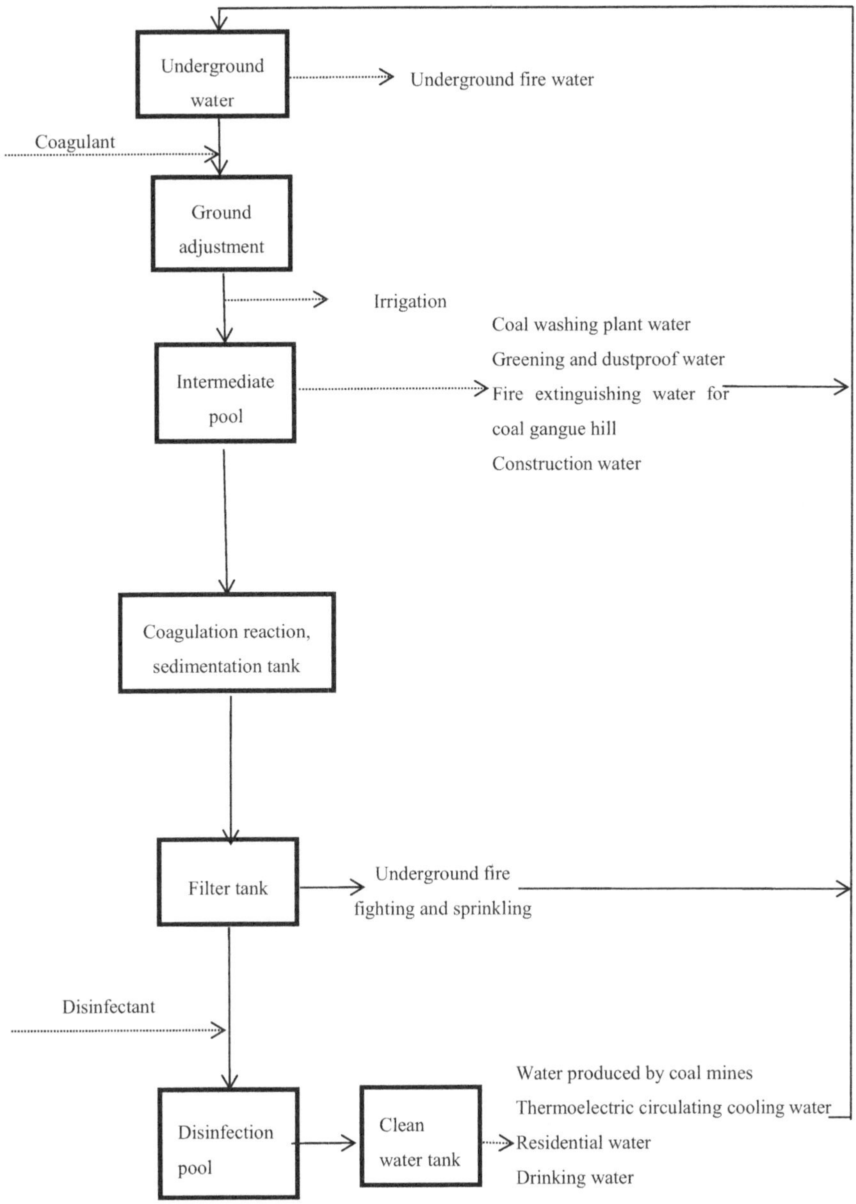

Fig. 5 Circular economy model of water resources in mining **areas**

(5) Prospecting technology around the mine.
(6) Research on alternative products.
(7) Pollution prevention technology.
(8) Geological disaster prevention technology.

4. Measures requirements for land reclamation engineering

For comprehensive reclamation and utilization of the land occupied, damaged, and reclaimed by mines, the reclamation quality shall comply with the provisions of the *Quality Control Standard for Land Reclamation* (TD/T 1036–2013), including the following aspects.

4.1 **Open-pit (pit)**

The shallow stope with a depth less than 1.0 m without waterlogging can be reclaimed into cultivated land by cutting the height to cushion the depression when the runoff conditions of surface water and groundwater are satisfied under natural conditions or after artificial restoration. The thickness of the covering soil depends on the weathering of the rock-soil at the pit′s bottom: When the weathering degree is high, the thickness of the natural settling soil is above 30 cm; when the rock-soil is complete with low weathering degree, the thickness is above 50 cm. Loam is the best overbarden, which can ensure the supply capacity of soil moisture. Deep excavation of open-pit mines without waterlogging will damage the land. Deep stopes with thin or thick overbardens, and shallow stopes with thick overbardens should be reclaimed into forest land.

The thickness of the covering soil, the configuration mode, and the planting method depend on the pit's bottom terrain, the weathering degree of the rock mass, the type of planted trees, and the root system's development. When the bottom terrain of the pit is flat, and the rock mass is seriously weathered, the whole soil should be covered, with the thickness of the natural settling soil above 30 cm; when the bottom terrain of the pit is undulating, and the rock mass is complete, the borrowed soil should be planted to reduce the treatment cost with earthwork reduction. Soil environmental quality should meet the requirements of *Soil Environmental Quality Risk Control Standard of Agricultural Land Soil Pollution (Trial)*" (GB 15,618–2018).

Open-pits with shallow water can be reclaimed for fishery (including aquaculture) by deep excavating and building small reservoirs; if near the town, it can be reclaimed into artificial waters and parks; if the accumulated water is above 3 m, it can be reclaimed into fishery (including aquaculture) or artificial waters and parks. The water quality of fishery (including aquaculture) should meet the *Water Quality Standard for Fishery* (GB 11,607–1989).

When the open-pit is used for construction land, its geological environment should be investigated to determine the development degree of adverse geological conditions such as collapse, landslide, fault and karst, and the bearing capacity indicators deformation, and stability of the foundation.

4.2 Subsided land

Reclamation of subsided land and eco-environment management should adjust measures to local conditions, retain the water surface of the waterlogged subsided land, and excavate reservoirs, impounding reservoirs, fishponds, or artificial lakes with techniques of digging deep to fill shallow and backfilling. The reclaimed water quality shall meet the water standards of Categories IV and V in the *Environmental Quality Standards for Surface Water* (GB 3838–2002).

Subsidence land with seasonal waterlogging, and local or seasonal waterlogged areas should be reclaimed into cultivated land, woodland, and grassland after shaping according to local conditions.

Local subsidence of hilly landforms formed by non-waterlogged subsided land, basically no water-logged or arid zones can be filled up. Slopes with a slope of less than 25° formed after the subsidence can be modified into horizontal terraces, while local small waterlogging can be transformed into paddy fields.

The filling of mine wastes shall refer to the national environmental standards, and the site shall be stabilized after sanitary safety landfilling. Besides, there shall be prevention and control measures to prevent the harmful components of the filling from polluting the groundwater and soil. In addition to compaction and other reinforcement measures, anti-seepage and anti-pollution treatments should be taken according to the nature and type of the fillers, and a liner isolation layer should be set if necessary.

4.3 Dumping site

The final slope (26 to 28°) of the dumping site should be adapted to the land use mode. The mechanical work area's slope is less than 20°, while the slope for ecological use is less than the natural repose angle (36°) of the rock-soil.

It is necessary to arrange the disposal order of rock-soil reasonably. Rock-soil containing undesirable ingredients shall be stacked in the deep; soil layers of suitable quality be stacked in the shallow, including easy-weathering terrane; soil layers rich in nutrients be stacked at the top or surface. The topsoil covering the surface can be collected before construction. Without suitable topsoil, it can be covered with other non-polluting materials confirmed by tests. The site use determines the thickness of the overburden.

If the mining stripping material contains toxic and harmful components, it should be covered deeply with gravel with no exposure, and anti-seepage measures should be taken. After covering the soil layer, it can be reclaimed as cultivated land.

The dumping site's supporting facilities should have a reasonable road layout, while the drainage facilities should meet the site requirements. Besides, there are measures to control soil erosion (especially for slopes) in design and construction.

4.4 Waste-rock yard

The newly discharged waste-rock should be compacted and rectified immediately to form a large-area and stable-slope reclamation site.

If the weathered layer is thick more than 10 cm, with fine particles and moderate pH, vegetation can be directly planted for reclamation without covering the soil.

However, if the weathered layer is thin, with high salt and acid pollution, it should be covered with soil over 30 cm after the pH is adjusted to a moderate value. The covering soil′s thickness of the non-weathered waste-rock should be above 50 cm.

When there is heavy metal pollution, if it is reclaimed as agricultural land, an isolation layer should be laid, and the soil should be over 50 cm.

The waste-rock yard's supporting facilities should have a reasonable road layout, while the drainage facilities should meet the site requirements. Besides, there are measures to control soil erosion, especially the soil erosion on slopes, in design and construction.

4.5 Tailings pond, red mud pile

If the engineering facilities cannot meet seepage and flood control requirements, technical engineering measures shall be taken.

The characteristics of wastes determine the necessity of covering soil and the thickness of the overbarden. With the covering-soil thickness above 50 cm, the covering-soil area should control soil and water loss.

Acidic, alkaline, or toxic and harmful pollutants determine the necessity, layer thickness, and material of the isolated layer according to the pollution content to ensure in-depth coverage.

The slope reclamation of red mud piles should select suitable vegetation substrates, construction techniques, and plant varieties based on the steep slope, high alkalinity, and difficulties for vegetation.

The manufacturer's authorised representative in the EU is Springer
Nature Customer Service Centre GmbH, Europaplatz 3, 69115 Heidelberg,
Germany. If you have any concerns regarding our products, please
contact ProductSafety@springernature.com

Printed and bound by CPI Group (UK) Ltd, Croydon, CR0 4YY
30/09/2026
02225505-0014